# MAGNETOHYDRODYNAMIC SHOCK WAVES

PUBLISHED 1963 BY THE M.I.T. PRESS, CAMBRIDGE, MASSACHUSETTS

J. EDWARD ANDERSON

Library of Congress Catalog Card Number: 62-22019

Printed in the United States of America

## FOREWORD

There has long been a need in science and engineering for systematic publication of research studies larger in scope than a journal article but less ambitious than a finished book. Much valuable work of this kind is now published only in a semiprivate way, perhaps as a laboratory report, and so may not find its proper place in the literature of the field. The present contribution is the sixteenth of the M.I.T. Press Research Monographs, which we hope will make selected timely and important research studies readily accessible to libraries and to the independent worker.

J. A. Stratton

To my wife, Naomi

# PREFACE

This monograph deals primarily with the existence, uniqueness, and qualitative properties of steady-state magnetohydrodynamic shock waves and with their stability with respect to breakup as a result of small-flow disturbances. As some knowledge of magnetohydrodynamic shock waves is assumed in the reader, the conservation (Rankine-Hugoniot) relations for these shock waves are discussed only in enough detail to provide background needed for the primary topic.

The main body of this monograph was submitted as a thesis to the faculty of the Department of Aeronautics and Astronautics at M. I. T. in partial fulfillment of the requirements for the degree of Doctor of Philosophy. While engaged in the thesis research, I had frequent discussions with Dr. William H. Heiser of the Mechanical Engineering Department at M. I. T., who at that time was preparing a magnetically driven shock tube for experiments on magnetohydrodynamic shock waves for his doctoral dissertation. These discussions, which were very helpful to me, led him to look for normal shock waves in the presence of a normal magnetic field of such a strength that the flow is super-Alfvénic ahead of the shock and sub-Alfvénic behind. The theory predicts that a shock wave of this type has a steady-state structure, that it can separate into a "switch-on" shock followed by a "switch-off" shock, but that the latter two shocks cannot retain their original form in the presence of normal Alfvén disturbances. After my monograph had been typed in final form, Heiser announced evidence that he had observed "switch-on" shocks. Because of the fundamental importance of these results to the whole question of existence and stability of magnetohydrodynamic shock waves, and the impetus they provide for further research, they are discussed by Dr. Heiser in the appendix to this monograph.

During my period of graduate study at M. I. T., I acquired more debts of gratitude than can be adequately acknowledged here. To each who helped I want to express my sincere thanks and deep appreciation for making this monograph possible. Professor Leon Trilling, my thesis supervisor, suggested that magnetohydrodynamic shock stability was an important and unsettled question, provided much encouragement, gave well-considered advice and criticism whenever needed, and helped in many other ways. Dr. Eugene E. Covert, a member of my doctoral committee, listened patiently through many discussions of progress and made many valuable suggestions. Professor Ascher Shapiro, the third member of my doctoral committee and also Dr. Heiser's thesis supervisor, contributed much to the success of this work by insisting that the theoretician keep in close touch with the real world and that the experimentalist be well versed in the theory. Professor J. Arthur Shercliff provided many helpful discussions during his year at M. I. T. and subsequently aided greatly by his terse, detailed criticisms of the manuscript. Professor Paul Germain and Dr. Heiser, as well as those mentioned above, read the manuscript and contributed helpful criticisms.

The monetary support which made it possible for me to give full attention to study and research came from a Convair Fellowship, from the Minneapolis-Honeywell Regulator Company, and from M. I. T. through a Research Assistantship, the funds for which were provided by the U. S. Air Force Office of Scientific Research under Contract AF-49(638)-207 and Grant AFOSR-62-84.

It has been a pleasure to work with the people of the M. I. T. Press, who have been most cooperative in making corrections and through whose suggestions the wording in a number of places has been improved. My wife, Naomi R. Anderson, through much self-sacrifice provided a home environment conducive to concentrated study and typed the first draft of the manuscript. Finally, our children Candy and Jim, though young, understood the need to leave their dad alone during his study periods and hence contributed immeasurably.

J. Edward Anderson

CONTENTS

## SYMBOLS

An arrow over a letter indicates a vector; the letter itself without a subscript indicates its scalar magnitude. The subscripts x, y, z indicate vector components.

| | |
|---|---|
| $A_\ell, A_r$ | amplitude of diverging wave moving to the left, right |
| a | speed of sound, $(\partial p/\partial \rho)_s^{1/2}$ |
| $\vec{B}$ | magnetic field |
| $\vec{b}$ | velocity of Alfvén waves, $\vec{B}/\sqrt{\mu\rho}$ |
| c | speed of light |
| $c_f, c_s$ | speeds of fast and slow magnetoacoustic waves, respectively |
| $\vec{E}$ | electric field |
| e | charge of the electron |
| $F_x$ | normal-momentum flux (Equation 2.8) |
| $F_y$ | transverse-momentum flux (Equation 2.9) |
| $\bar{F}_y$ | $F_y B_x/\mu G^2$ |
| $F_\tau, F_V, F_T,$ $F_B, F_J$ | null-surface functions (Equations 5.60 - 5.64) |
| f | $M_{f_1}{}^2$ (page 26) |
| G | mass flux, $\rho u$ |
| H | stagnation enthalpy (Equation 2.10) |
| h | enthalpy per unit mass |
| i | $\sqrt{-1}$ |
| $\vec{J}$ | conduction current |
| $\vec{j}$ | total current, $\vec{J} + \rho_c \vec{v}$ |
| $\vec{k}$ | propagation vector of small-amplitude waves (Equation 3.31) |
| k | Boltzmann's constant |
| L | arbitrary length used in dimensional analysis |

| | |
|---|---|
| $M_A$ | Alfvén number of flow, $u/b_x$ |
| $M_a$ | $a/b_x$ |
| $M_y$ | $b_y/b_x$ |
| $M_f, M_s$ | $c_f/b_x$, $c_s/b_x$ |
| $m_1$ | viscosity coefficient, $(4\eta/3)+\zeta$ |
| $m_2$ | $\eta$ |
| $m_+, m_-$ | mass of positive ion and electron, respectively |
| $n$ | number of particles per unit volume |
| $P_{ik}$ | pressure tensor (Equation 4.12) |
| $p$ | scalar pressure |
| $p_s$ | $(\partial p/\partial s)_\rho$ |
| $\vec{Q}$ | heat flux (Equation 4.13) |
| $R$ | gas constant |
| $r$ | $v_p/\sqrt{ab}$ (Equation 3.76) |
| $SP_i$ | $i^{th}$ singular point |
| $s$ | entropy per unit mass |
| $s$ | $M_{s_1}{}^2$ (page 26) |
| $T$ | temperature |
| $t$ | time |
| $U$ | internal energy per unit mass |
| $u$ | x-component of bulk velocity of flow, $v_x$ |
| $\vec{v}$ | bulk velocity of flow |
| $v$ | y-component of bulk velocity of flow, $v_y$ |
| $v_E$ | velocity of propagation of wave energy (Equation 3.60) |
| $v_p$ | phase velocity (Equation 3.47) |
| $v_g$ | group velocity (Equation 3.53) |
| $w$ | z-component of bulk velocity of flow, $v_z$ |
| $x$ | coordinate normal to the plane of the steady-state shock wave |
| $y$ | coordinate in plane of steady shock wave; steady-state $\vec{B}$ lies in x-y plane |
| $z$ | coordinate orthogonal to both x and y |

| | |
|---|---|
| $\alpha$ | angle between x-axis and $\vec{k}$ (Figure 3.9) |
| $\beta$ | density ratio across shock wave (page 19) |
| $\beta_m$ | maximum density ratio $(\gamma + 1)/(\gamma - 1)$ |
| $\gamma$ | ratio of specific heats |
| $\delta_{ij}$ | 1 if $i = j$, but 0 if $i \neq j$ |
| $\epsilon$ | $10^7/4\pi c^2$ farads/meter in MKS units |
| $\epsilon_{ijk}$ | permutation symbols (page 34) |
| $\zeta(y, t)$ | displacement of shock front from steady-state position (Figure 3. 1) |
| $\zeta, \eta$ | bulk and shear viscosity, respectively (Equation 4. 12) |
| $\eta$ | $M_{A_2}{}^2 - 1$ (page 25) |
| $\Theta$ | angle between $\vec{B}$ and direction of propagation of either the steady-state shock wave or a small-amplitude wave |
| $\kappa$ | coefficient of thermal conductivity (Equation 4. 25) |
| $\kappa'$ | coefficient relating conduction current to heat flux (Equation 4. 25) |
| $\lambda$ | an eigenvalue |
| $\vec{\lambda}$ | the eigenvector corresponding to $\lambda$ |
| $\mu$ | $4\pi(10)^{-7}$ henry/meter in MKS units |
| $\nu_c$ | collision frequency (Equation 4. 62) |
| $\xi$ | $M_{A_1}{}^2 - 1$ (page 25) |
| $\rho$ | mass density |
| $\rho_c$ | charge density |
| | electrical conductivity (Equation 4. 29) |
| | specific volume, $1/\rho$ |
| $\tau^*$ | specific volume corresponding to normal Alfvén speed, $B_x{}^2/\mu G^2$ (page 7) |
| $\phi$ | angle between $\vec{v}_g$ and $\vec{B}$ |
| $\chi$ | a dimensionless parameter (Equation 3. 76) |
| $\psi$ | invariant angle defined in Figure 3.10) |
| $\omega$ | angular frequency of small oscillations in shock reference frame (Equation 3. 31) |

| | |
|---|---|
| $\omega_0$ | angular frequency of small oscillations in reference frame at rest in the fluid (Equation 3.37) |
| $\omega_c$ | cyclotron frequency (Equation 4.63) |
| $\omega_p$ | plasma frequency (Equation 4.64) |
| (AB) | a representation of the dot product of two vectors $\vec{A}$ and $\vec{B}$ |
| { } | used to indicate the difference of the enclosed quantity on the two sides of the shock wave |

## Chapter 1

# INTRODUCTION

On the basis of the laws of conservation of mass, momentum, and energy and Maxwell's electromagnetic equations, several types of discontinuities can exist in ideal electrically conducting fluids in the presence of magnetic fields.[1] The discontinuities characterized by the condition that both the mass flow and density change across them are different from zero are called shock waves. This monograph is primarily concerned with one basic question related to these so-called "magnetohydrodynamic" shock waves: Can one expect to find them in nature? This question is too broad and complex to be treated exhaustively at the present time; hence, to make the subject of this monograph tractable, we shall restrict it in two ways: (1) by considering the shock wave as a local phenomenon away from physical boundaries and (2) by considering only those shock waves lying in a specified region of the phase space of the temperature, density, and magnetic-field variables.

We take into account the first restriction by assuming that the steady-state shock wave lies in a plane of infinite extent, with the properties of the flow varying only in the direction normal to that plane. This assumption, which says essentially that the local radius of curvature of the shock wave is large compared with its thickness, restricts the discussion to local properties of shock waves and leaves out of account the problem of stability in the large.

The second restriction is defined by the following assumptions;

1. The density of the gas is high enough and the magnetic field low enough so that the shock wave is collision-dominated; i.e., the collision frequency is large compared with the cyclotron frequency of the electrons. Use of this assumption is justified by the fact that there is a range of shock waves of experimental interest for which it is valid. It is manifested in the fact that the gas pressure and electrical conductivity are assumed to be scalar quantities, Hall currents are neglected, and the Navier-Stokes approximation is used for the pressure tensor and heat-flux vector.

2. The temperature and shock velocity are both low enough so that relativistic effects are unimportant, and the radiation

pressure and energy density are both small compared with the corresponding gas and magnetic quantities. These assumptions are justified both because shock waves in the broad region in which they are valid are not yet completely understood and because inclusion of these effects would make the theory considerably more difficult.

Moreover, in some of the discussions in Chapter 2 and in Chapters 5 and 6 it is assumed that the perfect-gas law holds. The reason for this assumption is that it enables one to visualize the form of certain integral surfaces and thus simplifies the proofs of the existence of shock waves. These existence proofs are of such a nature, however, that there is good reason to believe that the results will be valid for a wider range of gases, and in any case, an analytical framework has been established which will simplify the analysis of shock waves in real gases.

The study of magnetohydrodynamic shock waves was begun in 1950 with the paper of de Hoffmann and Teller.[2] Since then, continued interest inspired by astrophysics, by the possibilities of thermonuclear power, by flight at the outer edges of the atmosphere, etc., has produced many papers describing shock wave properties. The basic properties of magnetohydrodynamic shock waves as determined by the conservation laws (the Rankine-Hugoniot relations) have been developed further by Friedrichs,[3] Helfer,[4] Lüst,[5,6] Bazer and Ericson,[7] Napolitano,[8] and others, and they are now well understood; but the more complex question of their existence in nature has yet to be exhaustively treated.

The first efforts in this direction are due to several Russian authors, whose works are acknowledged and discussed in detail in Chapter 3. In their papers, the shock wave is considered to be a plane discontinuity in a perfect fluid, and the problem posed is to determine the stability of this configuration with respect to disintegration resulting from small disturbances in the flow. They have found quite simple criteria for the stability boundaries of shock waves and have also computed possible modes of disintegration of unstable shock waves. The main improvement that could be made on their work would be to provide greater physical insight into the conclusions reached.

There is another problem of fundamental importance to the present study: Can the nonlinear wave steepening processes which produce shock waves balance the diffusive processes in the fluid to such an extent that a steady-state shock wave will be maintained? This is the problem of the existence of the steady-state shock layer. It has had a more international history, reviewed at the beginning of Chapter 4, and should really come before study of the effects of small disturbances on shock waves. It is treated second in this monograph only because it is more difficult, requiring much more detailed knowledge of the equations of the flow, since all dissipative effects must be included. It has been found, in fact, that the basic equations of magnetohydrodynamics must be rederived for the shock-layer problem because the occurrence of finite gradients over distances of the order of the mean free path causes certain terms, neglected in the derivation of the usual macroscopic equations of magnetohydrodynamics, to become important. All previous papers on collision-dominated magnetohydrodynamic shock waves, to the author's knowledge, have used the latter equations; hence, the results are open to question.

The problems treated in the present monograph, as mentioned above, are: (1) the stability of shock waves considered as discontinuities in a perfect fluid and (2) the existence of the steady-state shock layer. It must be mentioned that both of these problems have been solved for ordinary gas dynamics, and the solutions have furnished essential background material and inspiration for the various attacks on the corresponding problems in magnetohydrodynamics. These problems are discussed in the introductions to Chapters 3 and 4, respectively, and also, for example, by Hayes.[9] It has been found by Gilbarg,[10] and in more detail by Gilbarg and Paolucci,[11] that normal gas-dynamic shock waves have a steady-state structure if and only if the flow upstream of the shock wave is supersonic and the flow downstream is subsonic. Under the same conditions, it has been found by Burgers[12] that these shock waves are also stable with respect to small disturbances in the flow. The situation in magnetohydrodynamics is far more complex; it is found there, for example, that certain shock waves which possess steady-state structure are not stable with respect to arbitrary small disturbances.

At each stage of development of the theory of this monograph, the basic equations are given only in the generality required at that stage, in order to avoid needless complication. Throughout this study, MKS units are used.

In Chapter 2, the magnetohydrodynamic Rankine-Hugoniot relations are briefly reviewed under the assumption that the reader has some familiarity with magnetohydrodynamic shock waves, such as can be obtained, for example, from Reference 1. The development leans toward those concepts which will be useful in later chapters, and has two novel features. One is the representation of the end states of shock waves as the intersection of three surfaces in the phase space of temperature, specific volume, and the component of the magnetic field parallel to the plane of the shock wave; the other is a derivation and presentation of the shock-wave adiabatic curve in a particularly illuminating form.

In Chapter 3, the existing literature on the problem of the reaction of magnetohydrodynamic shock waves to arbitrary small-flow perturbations is reviewed, and the ideas contained therein are somewhat augmented. This is the first problem mentioned above, and in it the shock wave is treated as a plane discontinuity in an infinite domain of perfect fluid. The concept of group velocity, essential in one part of the proof, is discussed exhaustively for magnetohydrodynamic and magnetoacoustic waves, and some arguments which give insight into the physical causes of instability of certain shock waves are presented.

In Chapter 4, a new derivation of the macroscopic equations of magnetohydrodynamics, valid for collision-dominated non-relativistic shock waves, is presented. These equations, based on the kinetic theory of fully ionized gases, have been closed by expressing the pressure tensor and heat-flux vector in terms of lower-order dependent variables by use of the phenomenological Navier-Stokes approximation. In the heat-flux vector, the cross-coupling effect of electromagnetic forces is included. By means of a dimensional analysis of the one-dimensional steady-state equations, it is deduced that the current-inertia terms in the generalized Ohm's law and the electric-force term in the momentum equation are by no means negligible within the shock wave. Each of these terms can cause important modifications in the structure of the shock wave.

In Chapters 5 and 6, the existence and uniqueness properties of shock waves are deduced. The analysis of these chapters is based for the most part on the work of Germain;[13] the inclusion of the current-inertia effect, however, has made the present analysis considerably more complex, but at the same time more nearly correct. Germain found that the methods he and others had used to prove existence and uniqueness of the shock layer were not sufficiently powerful to say much in general about slow and intermediate shock waves; inclusion of the current-inertia effect has made those methods even less conclusive. It has been found, however, that there are powerful arguments related to the topological behavior of integral curves between singular points which can be used to arrive at positive conclusions in all cases. In Chapter 5, the existence and uniqueness properties of shock waves which satisfy the equations derived in Chapter 4 are studied without making further approximations. The problem presented there is five-dimensional and as a result is difficult to visualize; hence, a number of reduced cases in two and three dimensions are solved in Chapter 6 to obtain a greater understanding of the topological behavior of the integral curves of the five-dimensional problem and also to study the effects of current inertia in a simpler context. Moreover, the special case in which the magnetic field has no component normal to the plane of the shock wave, and the cases of "switch-on" and "switch-off" shock waves, are given special consideration. As a by-product of the above analysis, the qualitative profile of the shock layer becomes clear, and a formula for the thickness of the shock layer results. The shock thickness based upon this formula is not sensitive to local variations in the shock profile and can be calculated numerically with relative ease.

In Chapter 7, results and conclusions obtained from the entire study are discussed. Several suggestions are made for further theoretical work and for means by which experimental confirmation of the principal results of the theory can be obtained.

Chapter 2

# SHOCK DISCONTINUITIES IN A PERFECT FLUID

## 1. The Basic Shock Relations

The relationships between the thermodynamic state, velocity, and magnetic field far upstream and far downstream of a shock wave (the Rankine-Hugoniot equations) are found by application of the laws of conservation of mass, momentum, and energy, and from the requirements of Maxwell's equations that the tangential component of the electric field and the normal component of the magnetic field are continuous across a discontinuity. Because the region of interest in this chapter is not the internal structure of the shock wave but the two regions, separated by the shock wave, in which the dissipation-producing gradients have vanished, the shock wave can be considered as a mathematical discontinuity in a perfect fluid.

It is well known[1] that a coordinate system can be chosen such that the velocity vectors and magnetic-field vectors on both sides of a magnetohydrodynamic shock wave lie in the same plane. If we take this plane to be the x-y plane, and assume the shock wave to lie in the y-z plane, the shock relations are

mass flux: $$\{\rho u\} = 0 \tag{2.1}$$

normal-momentum flux: $$\left\{p + \rho u^2 + \frac{B_y{}^2}{2\mu}\right\} = 0 \tag{2.2}$$

transverse-momentum flux: $$\left\{\rho u v - \frac{B_x B_y}{\mu}\right\} = 0 \tag{2.3}$$

energy flux: $$\left\{\rho u\left(h + \frac{u^2 + v^2}{2}\right) + \left(\vec{E} \times \frac{\vec{B}}{\mu}\right)_x\right\} = 0 \tag{2.4}$$

transverse electric field: $$\{\vec{E}_t\} = 0 \tag{2.5}$$

normal magnetic field: $$\{B_x\} = 0 \tag{2.6}$$

The braces $\{\ \}$ indicate the difference of the enclosed quantity on the two sides of the shock.

Assumption of a scalar pressure in Equation 2.2 implies that the case in which the cyclotron frequency is much smaller than the collision frequency is being considered. In the contrary case, since the pressure parallel to the magnetic field $p_{\parallel}$ is different from the pressure perpendicular to the magnetic field $p_{\perp}$, the pressure must be treated as a second-order tensor. Transformation of this tensor to x-y coordinates produces an off-diagonal term $p_{xy} = \frac{1}{2}(p_{\parallel} - p_{\perp}) \sin 2\theta$ which would appear in Equation 2.3. Thus, the usual magnetohydrodynamic shock relations derived here must be modified for low-density and/or high-magnetic-field situations.

We can further simplify the shock relations by moving the reference frame along the y-axis at a velocity such that either the transverse-momentum flux $F_y$ is zero, or the transverse electric field $\vec{E}_t$ is zero. (The latter alternative is possible only if $B_x \neq 0$; however, the case $B_x = 0$ can be treated as a relatively simple special case.) For the present chapter the choice makes no difference in the complexity of the resulting equations; but in Chapter 4 many more terms drop out if we take the second alternative; hence we shall make the assumption that $\vec{E}_t = 0$. The latter choice also has the conceptual advantage that the velocity and magnetic-field vectors are parallel on both sides of the shock. Since the current density must vanish outside the shock, $\vec{E} + \vec{v} \times \vec{B} = 0$ there; then, using $\vec{E}_t = 0$, Equation 2.5 can be replaced by the relation

$$u B_y - v B_x = 0 \tag{2.7}$$

on both sides of the shock.

Let $\rho u = u/\tau = G$, in which $\tau$ is the specific volume and $G$ is the mass flux, $B_x^2/\mu G^2 = \tau^*$, and eliminate $v$ by means of Equation 2.7. Then the shock relations become

$$p + G^2\tau + \frac{B_y^2}{2\mu} = F_x = \text{const} \tag{2.8}$$

$$\frac{B_y}{\mu}(\tau - \tau^*) = \bar{F}_y = \frac{F_y B_x}{\mu G^2} = \text{const} \tag{2.9}$$

$$h + \tfrac{1}{2} G^2 \tau^2 + \frac{\tau^2 B_y^2}{2\mu\tau^*} = H = \text{const} \tag{2.10}$$

in which $G$ and $\tau^*$ are constant and $F_x$ is the normal-momentum flux, $F_y$ the transverse-momentum flux, and H the stagnation enthalpy. Using the definitions of G and $\tau^*$, it is evident that a velocity $u^*$ can be defined by $u^* = G\tau^* = B_x^2/\mu G$ or

$$u^{*2} = \frac{B_x^2}{\mu}\tau^* = b_x^2$$

Thus, because of the proportionality between u and $\tau$, $\tau^*$ locates the Alfvén velocity on the $\tau$-axis.

## 2. Intersection of the Momentum and Energy Surfaces

In previous work, various types of plots of the shock relations have been presented in two dimensions; however, Equations 2.8, 2.9, and 2.10 show that the desired graphical representation is three-dimensional in two thermodynamic variables and $B_y$. Therefore only in a three-dimensional picture can all of the features of magnetohydrodynamic shocks be clearly visualized.

Equations of Surfaces to be Portrayed. We can make specific graphical representation only if we specify an equation of state; hence, for present purposes we assume the equation of state of an ideal gas. Substituting

$$p = \frac{RT}{\tau} \qquad \text{and} \qquad h = \frac{\gamma}{\gamma-1} RT$$

into Equations 2.8 and 2.10, the momentum and energy surfaces in RT - $\tau$ - $B_y$ space become

$$RT = \left(F_x - \frac{B_y^2}{2\mu}\right)\tau - G^2\tau^2 \qquad (2.8a)$$

$$\frac{B_y}{\mu}(\tau - \tau^*) = \bar{F}_y \qquad (2.9)$$

$$\frac{\gamma}{\gamma-1} RT = H - \tfrac{1}{2}G^2\tau^2 - \frac{\tau^2 B_y^2}{2\mu\tau^*} \qquad (2.10a)$$

If we change the sign of $\bar{F}_y$, the surface (Equation 2.9) is merely reflected in the plane $B_y = 0$. Then, since Equations 2.8a and 2.10a are symmetrical about $B_y = 0$, we can assume without loss of generality that $\bar{F}_y \geq 0$.

Normal-Momentum Surface. Equation 2.8a represents a parabola in all planes of intersection in which either $B_y$ or $\tau$ is constant, and is symmetric about the plane $B_y = 0$. When $B_y = 0$, Equation 2.8a intersects the $\tau$-axis at the points $\tau = 0$, $F_x/G^2$. Its maximum is at $\tau = F_x/2G^2$, $RT = (F_x/2G)^2$. When $RT = 0$, either $\tau = 0$ or $G^2\tau = F_x - (B_y^2/2\mu)$. When $\tau = 0$ in this parabola, $B_y = \pm\sqrt{2\mu F_x}$. With these facts, the $F_x$ surface can be drawn as shown in Figure 2.1.

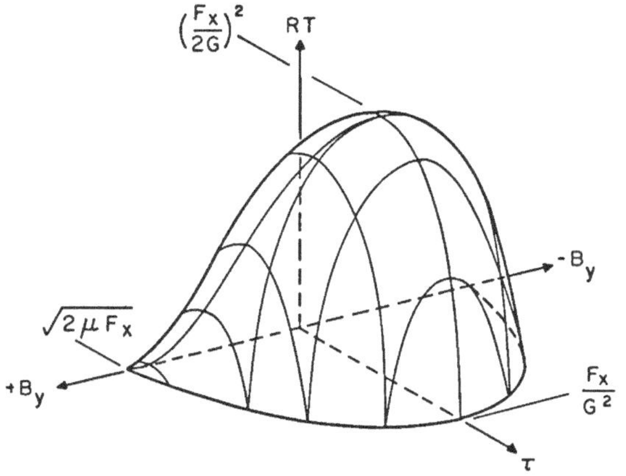


Figure 2.1. Surface of constant normal-momentum flux

Transverse-Momentum Surface. Equation 2.9 represents a hyperbolic cylinder asymptotic to $\tau = \tau^*$ with the generating lines parallel to the RT-axis. When $\bar{F}_y > 0$, it is as shown in Figure 2.2.

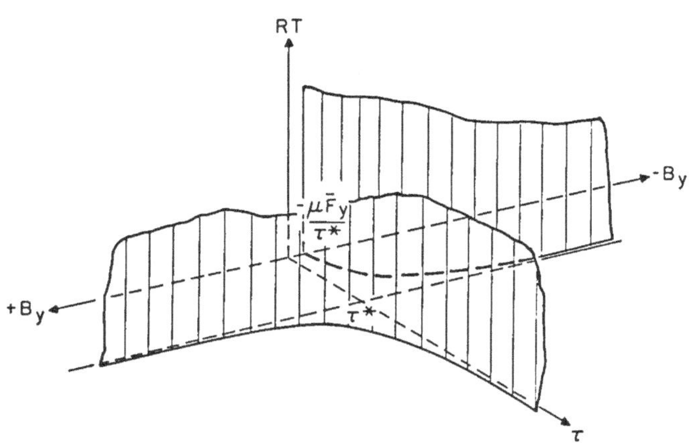


Figure 2.2. Surface of constant transverse-momentum flux

Energy Surface. Equation 2.10a also represents an unbounded surface in $RT > 0$, $\tau > 0$. All sections normal to the $\tau$- and $B_y$-axes are parabolic, and the surface is symmetric about both of these axes. In the plane $B_y = 0$, the parabola intersects the $\tau$-axis at $(2H)^{1/2}/G$ and the $RT$-axis at $[(\gamma - 1)/\gamma]\, H$. In the plane $\tau = 0$, the surface becomes a straight line $RT = [(\gamma - 1)/\gamma]\, H$, and in the plane $RT = 0$ it is a symmetric hyperbolic curve asymptotic to the $\pm B_y$-axes with inflection points at $B_y = \pm G(\mu\tau^*/2)^{1/2}$. Using this information, we may sketch the energy surface as in Figure 2.3.

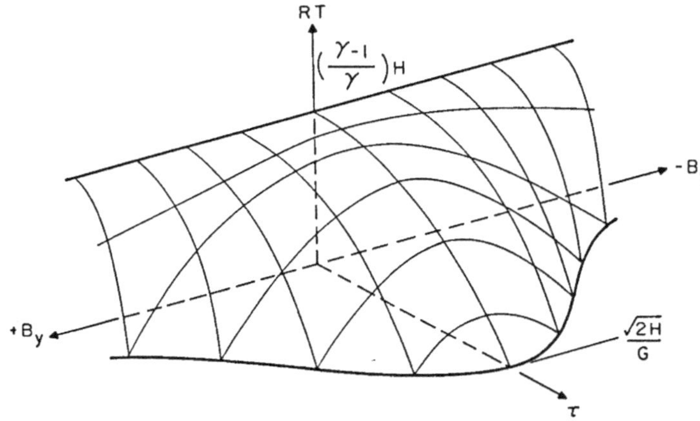


Figure 2.3. Surface of constant stagnation enthalpy

Intersections between the Normal-Momentum and Energy Surfaces. Intersection of the normal-momentum surface with the energy surface gives a space curve which can be visualized qualitatively from Figures 2.1 and 2.3. Intersection of this curve with the transverse-momentum surface then gives the points which may possibly be connected by shocks. We can see the analytical properties of the former intersection if we eliminate $RT$ between Equations 2.8a and 2.10a. The result solved for $B_y^2/2\mu$ is

$$\frac{B_y^2}{2\mu} = \frac{\frac{1}{2}\left(\frac{\gamma + 1}{\gamma - 1}\right) G^2\tau^2 - \frac{\gamma}{\gamma - 1} F_x\tau + H}{\frac{\tau}{\tau^*}\left(\tau - \frac{\gamma}{\gamma - 1}\,\tau^*\right)} \qquad (2.11)$$

By noting that the curve of intersection exists only for values of $\tau$ for which the right-hand side of Equation 2.11 is positive, and observing the location of the roots ($\tau_1$, $\tau_2$) of the numerator, we can readily plot Equation 2.11. Four cases can be distinguished:

1. $\tau_1$, $\tau_2$ real and $\tau_1 < \tau_2 < \frac{\gamma}{\gamma - 1}\tau^*$

2. $\tau_1$, $\tau_2$ real and $\tau_1 < \frac{\gamma}{\gamma - 1}\tau^* < \tau_2$

3. $\tau_1$, $\tau_2$ real and $\frac{\gamma}{\gamma - 1}\tau^* < \tau_1 < \tau_2$

4. $\tau_1$, $\tau_2$ complex conjugates

In Figure 2.4, Equation 2.11 is plotted qualitatively for each of these cases.

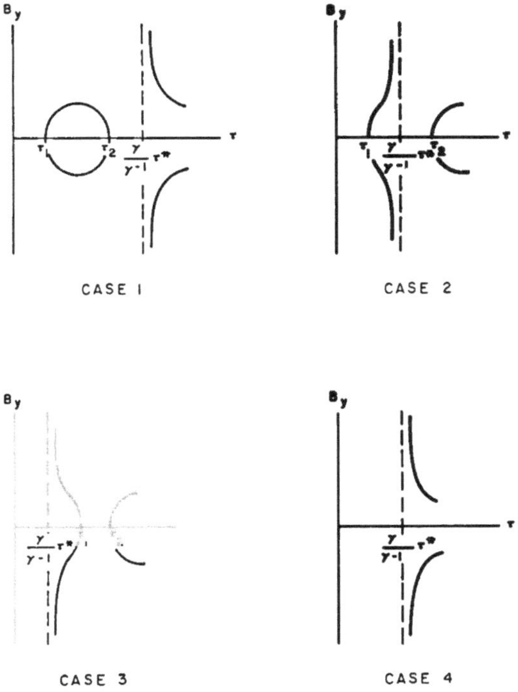


Figure 2.4. Projection on the $B_y$-$\tau$ plane of the intersection of the normal-momentum flux and energy surfaces

The interpretation of the points $\tau_1$ and $\tau_2$ is that they are the values of $\tau$ downstream and upstream of a shock with $B_y = 0$, that is, in ordinary hydrodynamics. In case 4, therefore, shocks cannot exist for which $B_y = 0$ on either or both sides, and in the following paragraph we show that for this case there are no shocks.

Intersection with the Transverse-Momentum Surface. Points which may possibly be connected by shocks can be visualized by superimposing Figure 2.2 upon Figure 2.4. Since $[\gamma/(\gamma - 1)] > 1$, it is immediately obvious that in case 4 there can be no more than one intersection; hence, no shocks are possible, and a requirement for the existence of shocks is that the discriminant of the numerator of Equation 2.11 must be positive. In case 3, there will be only two intersections, both for $\tau > \tau^*$. In case 2, there are two intersections in the region for $\tau > \tau^*$, and there are two or zero intersections for $\tau < \tau^*$. In case 1, there can be four intersections if and only if $\tau_1 < \tau^* < \tau_2$.

It is clear that there are at most four points of intersection which may possibly be connected by shocks, and that all four can occur only in cases 1 and 2. (As shown on page 15, the existence of at most four points of intersection does not depend on the perfect-gas assumption.) The intersections corresponding to these cases are depicted three-dimensionally in Figures 2.5 and 2.6, respectively. The points numbered 1, 2, 3, 4, ordered accord-

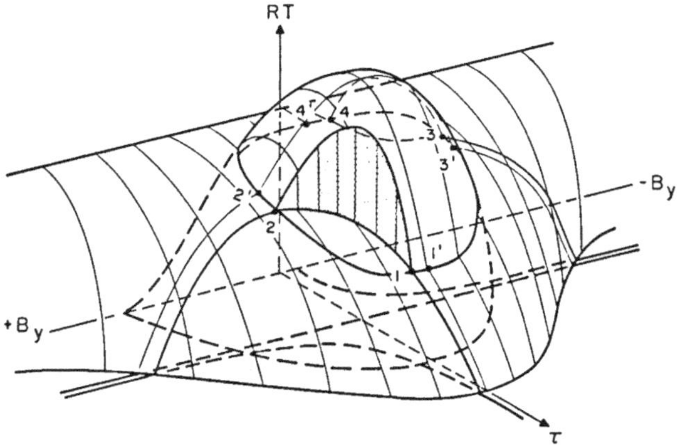


Figure 2.5. Intersection of surfaces in Figures 2.1, 2.2, and 2.3 for case 1 of Figure 2.4

ing to decreasing specific volume, are the points at which all three surfaces intersect and thus are possible end points for shocks. (The designation used here for these points is that instituted by Germain.[13]) Points 2' and 3' are the points where the plane $\tau = \tau^*$ intersects the curve of intersection of the $F_x$ and H surfaces, and points 1' and 4' are the points where the plane $B_y = 0$ intersects the same curve. Equation 2.9 and Figure 2.2 show that as $\overline{F}_y$ goes to zero, points 1 and 2 approach 1' and 2', respectively; and points 3 and 4 approach 3' and 4', respectively. These limiting cases are referred to as "switch-on" and

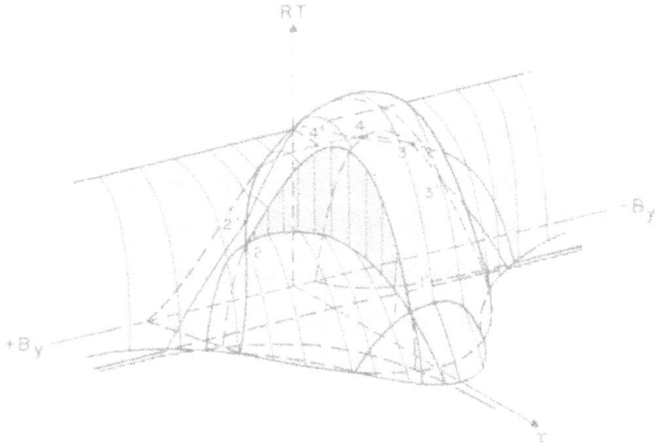


Figure 2.6. Intersection of surfaces in Figures 2.1, 2.2, and 2.3 for case 2 of Figure 2.4

"switch-off" shocks.

Since $u = G\tau$, the $\tau$-axis can also be considered as the u-axis, and since $\tau^*$ corresponds to the Alfvén speed, it is clear that fast shocks ($1 \rightarrow 2$) are super-Alfvénic, and slow shocks ($3 \rightarrow 4$) are sub-Alfvénic. The arrows indicate the direction of the transitions for compressive shocks; the next paragraph demonstrates that entropy increases in the same direction. With that in mind one can see directly that temperature increases across both fast and slow shocks, and the magnetic field increases across fast shocks and decreases across slow shocks. As shown by Germain[13] and Shercliff,[14] the conservation laws also allow the "intermediate-shock" transitions $1 \rightarrow 3$, $1 \rightarrow 4$, $2 \rightarrow 3$, $2 \rightarrow 4$.

## 3. Ordering of the Shock Transition Points According to Increasing Entropy

Because of its importance in the study of existence of shocks, a proof is given here of the fact that points 1, 2, 3, 4 are ordered according to increasing entropy, that is,

$$s_1 \leq s_2 \leq s_3 \leq s_4$$

This theorem is proved by Germain,[13] and again by Shercliff[14] in a different way. In the following paragraphs, we shall use Shercliff's arguments in order to prove that $s_1 \leq s_2$ and $s_3 \leq s_4$, and an argument similar to Germain's to prove that $s_2 \leq s_3$.

Entropy Rise across Fast and Slow Shocks. If $B_y$ is eliminated from Equations 2.8 and 2.10 by substitution from Equation 2.9, the following two equations result:

$$p + G^2\tau + \frac{1}{2}\frac{\mu\overline{F}_y^{\,2}}{(\tau - \tau^*)^2} = F_x \qquad (2.12)$$

$$h + \frac{1}{2}G^2\tau^2 + \frac{\mu\tau^2}{2\tau^*}\frac{\overline{F}_y^{\,2}}{(\tau - \tau^*)^2} = H \qquad (2.13)$$

Equation 2.12, the Rayleigh line given by Germain,[13] has two branches in the plane $p > 0$, $\tau > 0$. These two branches are the heavy solid curves of Figure 2.7. The intersection of the Rayleigh line with Equation 2.13 gives the points which can be connected by shocks.

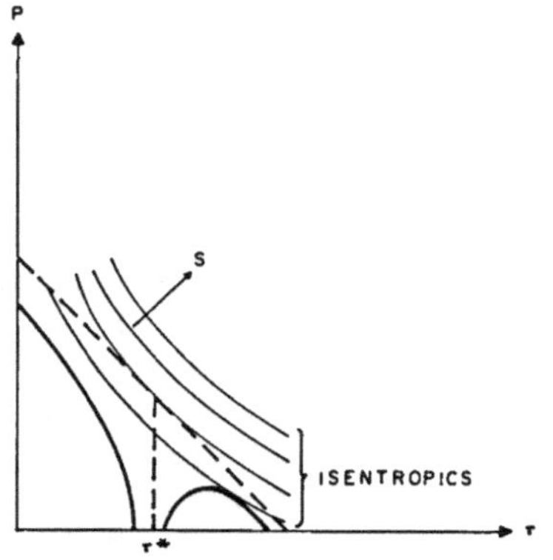


Figure 2.7. Magnetohydrodynamic Rayleigh line

Consider the family of curves in the $p$ - $\tau$ plane obtained from Equation 2.13 for all possible values of H. If we allow H to vary with $F_x$, $\overline{F}_y$, $\tau^*$, and G to remain fixed, we obtain the following differential equations from Equations 2.12 and 2.13:

$$dp + G^2 d\tau = \frac{\mu\overline{F}_y^{\,2}}{(\tau - \tau^*)^3}\,d\tau \qquad (2.14)$$

$$dh + G^2\tau\,d\tau = \frac{\mu\overline{F}_y^{\,2}}{(\tau - \tau^*)^3}\,\tau\,d\tau + dH \qquad (2.15)$$

Hence, using a familiar equation from thermodynamics,

$$T\,ds = dh - \tau\,dp = dH \qquad (2.16)$$

Thus, extrema of $s$ occur together with extrema of H, and since $\int_1^2 dH = 0$ across a possible shock transition,

$$\int_1^2 T\,ds = 0 \qquad (2.17)$$

along the Rayleigh line. Here, the suffixes 1 and 2 refer generally to the states ahead of and behind a shock.

Now Weyl[15] has stated the criterion that for all known gases the isentropic lines in the $p-\tau$ plane have negative slope and positive curvature. Using this criterion, we can see from Figure 2.7 that there can be only one entropy extremum point (a maximum) on each branch of the Rayleigh line. Then from Equation 2.16, it is evident that there is only one maximum value of H on each branch. If this is true, there can be at most two intersections of a given H-curve (Equation 2.13) with each branch of the Rayleigh line; hence, as found above for a perfect gas, there are at most four points which may be connected by shocks for a given set of the constants $F_x$, $F_y$, G, H, $B_x$.

At the point on the Rayleigh line where H is a maximum ($ds = 0$), Equation 2.12 gives

$$\left(\frac{\partial p}{\partial \tau}\right)_s = -G^2 + \frac{\mu \overline{F}_y^{\,2}}{(\tau - \tau^*)^3} \tag{2.18}$$

If the definition of the speed of sound, $a^2 = -\tau^2(\partial p/\partial \tau)_s$, the definitions of G, $\overline{F}_y$, and $\tau^*$, and Equation 2.7 are substituted into Equation 2.18, it becomes

$$u^4 - (b_x^{\,2} + b_y^{\,2} + a^2)\,u^2 + a^2 b_x^{\,2} = (u^2 - c_f^{\,2})(u^2 - c_s^{\,2}) = 0 \tag{2.19}$$

which is the well-known equation for the speeds of fast and slow small-amplitude magnetohydrodynamic waves. Thus, the points on the Rayleigh line corresponding to entropy maxima are the magnetosonic points, and, for a shock of finite strength, each pair of intersections of the Rayleigh line with the H-line must straddle these points. Taking into account the fact that the fast and slow shocks are separated by the Alfvén speed, the velocities at points 1, 2, 3, 4 can now be ordered as follows:

$$u_1 \geq c_f \geq u_2 \geq b_x \geq u_3 \geq c_s \geq u_4 \tag{2.20}$$

As pointed out by Shercliff,[14] Equation 2.17 suggests plotting the Rayleigh line in T-s coordinates. Remembering how the isentropic lines go in the $p-\tau$ plane, we can easily execute the qualitative shape of this plot; this is shown in Figure 2.8.

With s fixed, higher temperature corresponds to lower specific volume in a Weyl gas; hence, the ordering of the points 1, 2, 3, 4 is as shown in Figure 2.8. In order to satisfy Equation 2.17, which is integral along the Rayleigh line, the area under the Rayleigh line from point 1 to the sonic point must equal the area under the Rayleigh line from the sonic point to point 2. From this it is obvious that $s_1 \leq s_2$ and similarly $s_3 \leq s_4$. QED

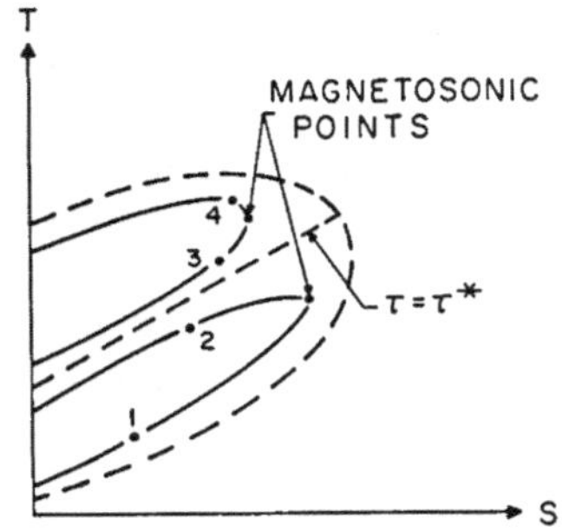

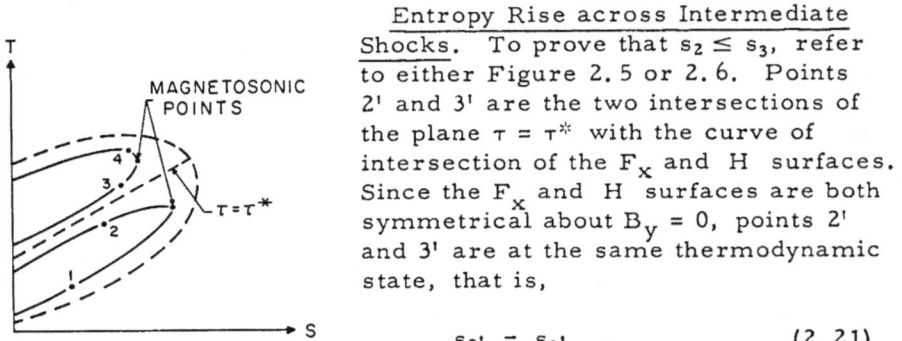

Figure 2.8. Magnetohydrodynamic Rayleigh line of Figure 2.7 plotted in T-s coordinates

Entropy Rise across Intermediate Shocks. To prove that $s_2 \leq s_3$, refer to either Figure 2.5 or 2.6. Points 2' and 3' are the two intersections of the plane $\tau = \tau^*$ with the curve of intersection of the $F_x$ and H surfaces. Since the $F_x$ and H surfaces are both symmetrical about $B_y = 0$, points 2' and 3' are at the same thermodynamic state, that is,

$$s_{2'} = s_{3'} \tag{2.21}$$

Along the segments 2-2' and 3-3', $F_x$ and H are constant, but $\overline{F}_y$ varies. Study of Equation 2.9 and Figure 2.2 reveals that $\overline{F}_y = 0$ at points 2' and 3', and increases monotonically up to its value at points 2 and 3. This also follows from the fact stated earlier that there are at most four points which can be connected by shocks, and hence, only one extremum point for $\overline{F}_y$ can exist on each side of the $B_y$-axis. Thus, we may consider the variation of Equations 2.8, 2.9, and 2.10 with $F_x$, H, G, $\tau^*$ fixed. Performing this variation,

$$dp + G^2 d\tau + \frac{B_y}{\mu} dB_y = 0 \tag{2.22}$$

$$\frac{B_y}{\mu} d\tau + \frac{\tau - \tau^*}{\mu} dB_y = d\overline{F}_y \tag{2.23}$$

$$dh + G^2 \tau d\tau + \frac{\tau^2 B_y}{\mu \tau^*} dB_y + \frac{B_y{}^2 \tau}{\mu \tau^*} d\tau = 0 \tag{2.24}$$

Hence,

$$T\, ds = dh - \tau dp = -\frac{\tau B_y}{\tau^*}\left[\frac{B_y}{\mu} d\tau + \frac{\tau - \tau^*}{\mu} dB_y\right]$$

$$= -\frac{\tau}{\tau^*} B_y\, d\overline{F}_y = f(\overline{F}_y)\, d\overline{F}_y \tag{2.25}$$

where for the transition $2 \rightarrow 2'$, $f(\overline{F}_y) < 0$, and for $3' \rightarrow 3$, $f(\overline{F}_y) > 0$. Since $f(\overline{F}_y)$ is a single-valued function,

$$\int_{2}^{2'} f(\bar{F}_y)\, d\bar{F}_y = \int_{0}^{F_{y_2}} [-f(\xi)]\, d\xi > 0$$

is a monotonically increasing function of $\bar{F}_y$. Since

$$\int_{2}^{2'} T ds = \int_{0}^{F_{y_2}} [-f(\xi)]\, d\xi$$

the integral on the left is also monotonically increasing from 2 to 2'. Since $T > 0$, this is possible only if

$$s_2 \leq s_{2'} \tag{2.26}$$

the equality applying only if 2 coincides with 2'. Similarly,

$$\int_{3'}^{3} f(\bar{F}_y)\, d\bar{F}_y = \int_{0}^{\bar{F}_{y_3}} f(\xi)\, d\xi > 0$$

so that

$$s_{3'} \leq s_3 \tag{2.27}$$

With the help of Equation 2.21, it is clear that the inequality

$$s_2 \leq s_3 \tag{2.28}$$

always holds. QED

Germain gives $(\partial h/\partial p)_T - 2\tau > 0$ as a sufficient but not a necessary condition that the above proof is valid. As he shows, the necessary condition, $(\partial h/\partial p)_T - 2\tau + \tau^* > 0$, is that $p(\tau)$ be monotone along the segments 2 - 2', 3 - 3'; in other words, that $f(\bar{F}_y)$ be monotone. It has already been proved that this is true.

## 4. The Shock Adiabatic

Figures 2.5 and 2.6 form one useful type of representation of magnetohydrodynamic-shock end states, which, for existence studies, is perhaps the most useful. For the stability studies of Chapter 3, however, graphs of the flow velocity downstream of the shock as a function of flow velocity upstream are more directly useful. Plots of this type are frequently referred to as shock adiabatics.

For oblique magnetohydrodynamic shocks, Polovin and Demutskii[16] have derived approximate relationships for the shock adiabatic for perfect gases by expanding the jumps in the various dependent variables in a power series in $\{\rho\}/\rho$ up to second order. In this section the exact equation describing the shock adiabatic for perfect gases is found, and it is shown that because of its erratic behavior in certain regions, the approximate solution is not very useful; in fact, the approximate formulas are actually more complex than the exact solution.

In the work of this section we convert all velocities to dimensionless form by dividing by the normal component of the Alfvén speed $b_x$. In this way we can obtain results of the greatest possible generality.

Derivation of the Exact Equation for the Shock Adiabatic. Using the perfect-gas relationships $p = \rho a^2/\gamma$ and $h = a^2/(\gamma - 1)$, Equations 2.1 through 2.6 may be written

$$\rho_1 u_1 = \rho_2 u_2 = G \tag{2.29}$$

$$\frac{\rho_1 a_1^2}{\gamma} + \rho_1 u_1^2 + \frac{B_{y_1}^2}{2\mu} = \frac{\rho_2 a_2^2}{\gamma} + \rho_2 u_2^2 + \frac{B_{y_2}^2}{2\mu} \tag{2.30}$$

$$\rho_1 u_1 v_1 - \frac{B_x B_{y_1}}{\mu} = \rho_2 u_2 v_2 - \frac{B_x B_{y_2}}{\mu} \tag{2.31}$$

$$\rho_1 u_1\left(\frac{a_1^2}{\gamma - 1} + \frac{u_1^2 + v_1^2}{2}\right) + \frac{B_{y_1}}{\mu}\left(u_1 B_{y_1} - v_1 B_x\right)$$

$$= \rho_2 u_2\left(\frac{a_2^2}{\gamma - 1} + \frac{u_2^2 + v_2^2}{2}\right) + \frac{B_{y_2}}{\mu}\left(u_2 B_{y_2} - v_2 B_x\right) \tag{2.32}$$

$$v_1 B_x - u_1 B_{y_1} = v_2 B_x - u_2 B_{y_2} \tag{2.33}$$

$$B_{x_1} = B_{x_2} = B_x \tag{2.34}$$

The first step toward the desired end results is to eliminate all quantities on side 2 except the density $\rho_2$. To do this, solve Equations 2.31 and 2.33 for $v_2$ and $B_{y_2}$, resulting in

$$v_2 = v_1 + \frac{(\beta - 1) M_{y_1}}{M_{A_1}^{\ 2} - \beta} u_1 \tag{2.35}$$

$$B_{y_2} = \frac{M_{A_1}^{\ 2} - 1}{\dfrac{M_{A_1}^{\ 2}}{\beta} - 1} B_{y_1} \tag{2.36}$$

in which the following dimensionless parameters are used:

$\beta = \rho_2 / \rho_1$ = density ratio

$M_A = u/b_x$ = Alfvén number of the flow

$M_y = b_y/b_x$ = $\tan \theta$

Also,

$M_a = a/b_x$ will be used below.

The following quantities, appearing in Equations 2.30 and 2.32, can be obtained from Equations 2.35 and 2.36:

$$\frac{B_{y_2}^{\ 2} - B_{y_1}^{\ 2}}{2\mu} = (\beta - 1) M_{A_1}^{\ 2} \frac{\left[(\beta + 1) M_{A_1}^{\ 2} - 2\beta\right]}{(M_{A_1}^{\ 2} - \beta)^2} \frac{B_{y_1}^{\ 2}}{2\mu} \tag{2.37}$$

$$\frac{B_{y_2}{}^2}{\mu \rho_2} \quad \frac{B_{y_1}{}^2}{\mu \rho_1} = (\beta \quad 1) \frac{(M_{A_1}{}^4 - \beta)}{(M_{A_1}{}^2 - \beta)^2} \frac{B_{y_1}{}^2}{\mu \rho_1} \tag{2.38}$$

$$\tfrac{1}{2}(v_2{}^2 - v_1{}^2) - \frac{B_x}{\mu G}(B_{y_2} v_2 - B_{y_1} v_1)$$

$$= -\frac{(\beta - 1)}{2} \frac{\left[(\beta + 1)M_{A_1}{}^2 - 2\beta\right]}{(M_{A_1}{}^2 - \beta)^2} \frac{B_{y_1}{}^2}{\mu \rho_1} \tag{2.39}$$

If we divide Equation 2.30 by $B_x{}^2/\mu$, it becomes dimensionless. If we substitute Equation 2.37 and use dimensionless parameter $M_a{}^2 = \mu \rho a^2 / B_x{}^2$, it becomes

$$1/\gamma (M_{a_2}{}^2 - M_{a_1}{}^2) + M_{A_2}{}^2 - M_{A_1}{}^2$$

$$+ (\beta - 1) M_{A_1}{}^2 \frac{\left[(\beta + 1)M_{A_1}{}^2 - 2\beta\right]}{(M_{A_1}{}^2 - \beta)^2} \frac{M_{y_1}{}^2}{2} = 0 \tag{2.40}$$

Dividing Equation 2.32 by Equation 2.29, and also by $B_x{}^2/\mu \rho_2$, and substituting Equations 2.38 and 2.39, we may write the result in the form

$$\frac{1}{\gamma - 1}(M_{a_2}{}^2 - M_{a_1}{}^2) \quad \frac{(\beta - 1)}{(\gamma - 1)} M_{a_1}{}^2 \quad \tfrac{1}{2}(M_{A_2}{}^2 - \beta M_{A_1}{}^2)$$

$$+ \tfrac{1}{2}\beta(\beta - 1) \frac{(2M_{A_1}{}^2 - \beta - 1)}{(M_{A_1}{}^2 - \beta)^2} M_{A_1}{}^2 M_{y_1}{}^2 = 0 \tag{2.41}$$

Now, since

$$\frac{M_{A_1}^{\ 2}}{M_{A_2}^{\ 2}} = \frac{u_1^{\ 2}}{\frac{B_x^{\ 2}}{\mu\rho_1}} \div \frac{u_2^{\ 2}}{\frac{B_x^{\ 2}}{\mu\rho_2}} = \left(\frac{\rho_1}{\rho_2}\right)\left(\frac{\rho_2}{\rho_1}\right)^2 = \beta \qquad (2.42)$$

the desired result may take two different forms, each having its own advantage. First we eliminate the quantity $M_{a_2}^{\ 2} - M_{a_1}^{\ 2}$ between Equations 2.40 and 2.41. Then, in the result, we eliminate $M_{A_2}^{\ 2}$ by means of Equation 2.42 to give the following expression for $M_{A_1}^{\ 2}$:

$$(M_{A_1}^{\ 2} - \beta)^2 \left\{ M_{A_1}^{\ 2} - \frac{2\beta M_{a_1}^{\ 2}}{\beta + 1 - \gamma(\beta - 1)} \right\}$$

$$- \beta M_{y_1}^{\ 2} \left\{ \left[\frac{2\beta - \gamma(\beta - 1)}{\beta + 1 - \gamma(\beta - 1)}\right] M_{A_1}^{\ 2} - \beta \right\} M_{A_1}^{\ 2} = 0 \qquad (2.43)$$

This is the exact equation for the shock adiabatic. The second form (Equation 2.58) is obtained by substituting for $\beta$ from Equation 2.42.

The Limit of Weak Shocks. Equation 2.43 is a bicubic equation for $M_{A_1}^{\ 2}$ in terms of the quantities $M_{a_1}^{\ 2}$, $M_{y_1}^{\ 2}$, and $\beta$, in which it should be noted that $M_{a_1}^{\ 2}$ represents the pressure on the upstream side of the shock while $M_{y_1}^{\ 2}$ determines the inclination of the magnetic field on the upstream side. Equation 2.43 clearly shows that three finite amplitude shocks are possible, corresponding to the slow, intermediate, and fast shocks discussed in Section 2. When $\beta = 1$, Equation 2.43 reduces to

$$(M_{A_1}^{\ 2} - 1)\left[(M_{A_1}^{\ 2} - 1)(M_{A_1}^{\ 2} - M_{a_1}^{\ 2}) - M_{y_1}^{\ 2} M_{A_1}^{\ 2}\right] = 0 \qquad (2.44)$$

which is recognized as the expression for the speeds of propagation of small disturbances in a perfectly conducting fluid situated in a magnetic field, the first factor representing Alfvén waves and the second representing magnetoacoustic waves.

Greater transparency is attainable in the following paragraphs if we express $M_{a_1}^{\ 2}$ and $M_{y_1}^{\ 2}$ in terms of the squares of the fast and slow sound speeds $M_{f_1}^{\ 2}$ and $M_{s_1}^{\ 2}$ because, as is shown by Inequalities 2.20, $M_{f_1}^{\ 2}$ and $M_{s_1}^{\ 2}$ are specifically oriented by the inequality

$$M_s \leq 1 \leq M_f \tag{2.45}$$

To be consistent with Equation 2.19, $M_f$ and $M_s$ must be defined by the following relation:

$$(M_A^2 - 1)(M_A^2 - M_a^2) - M_y^2 M_A^2 = (M_A^2 - M_f^2)(M_A^2 - M_s^2) \tag{2.46}$$

Equation 2.46 holds true in general since it is a definition, and in particular when $M_A^2 = M_f^2$ or $M_s^2$. Substituting each of these quantities for $M_A^2$ gives two equations:

$$\left.\begin{aligned} (M_f^2 - 1)M_a^2 + M_f^2 M_y^2 &= (M_f^2 - 1)M_f^2 \\ (M_s^2 - 1)M_a^2 + M_s^2 M_y^2 &= (M_s^2 - 1)M_s^2 \end{aligned}\right\} \tag{2.47}$$

which may be solved for $M_a^2$ and $M_y^2$, yielding

$$\left.\begin{aligned} M_a^2 &= M_f^2 M_s^2 \\ M_y^2 &= (M_f^2 - 1)(1 - M_s^2) \end{aligned}\right\} \tag{2.48}$$

or solving for $M_f^2$ and $M_s^2$, and manipulating the expression under the radical:

$$M_{f,s}^2 = \frac{1}{2}\left\{ M_a^2 + M_y^2 + 1 \pm \sqrt{(M_a^2 + M_y^2 - 1)^2 + 4M_y^2} \right\} \tag{2.49}$$

Another form clearly reveals the limits on $M_f^2$ and $M_s^2$. Set the left side of Equation 2.46 equal to zero and multiply by $b_x^4$. The result can be expressed in the form

$$\left(u^2 - \frac{a^2 + b^2}{2}\right)^2 + a^2 b^2 \cos^2\theta = \left(\frac{a^2 + b^2}{2}\right)^2 \tag{2.50}$$

in which $b^2 = b_x^2 + b_y^2$. This is the equation of a circle of radius $\frac{1}{2}(a^2 + b^2)$ in $u^2$, $ab\cos\theta$ coordinates, which is plotted in Figure 2.9. As $\theta$ ranges through all possible values, the fast and slow sound speeds range in the heavily outlined portions of the circle. The following simple calculation from the figure shows the range of fast and slow sound. In triangle ABC the sides AC and BC are known. Hence,

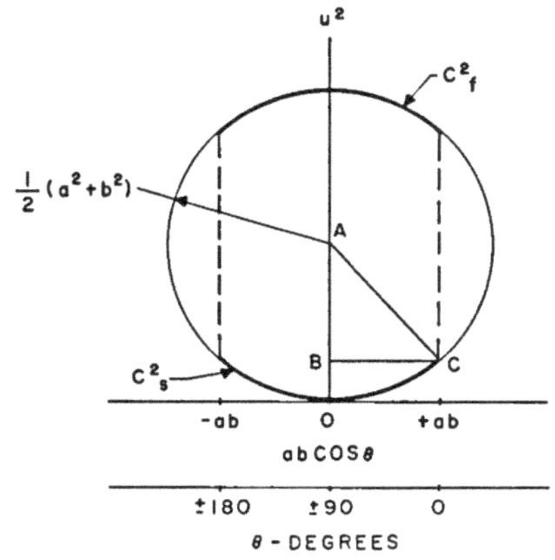


Figure 2.9. Locus of fast and slow sound speeds

$\overline{AB}$ $\frac{1}{2}|a^2 - b^2|$

and then

$$c_{s_{max}}{}^2 = \overline{BO}$$
$$= \tfrac{1}{2}(a^2 + b^2 - |a^2 \quad b^2|)$$
$$= \text{Min}\,(a^2, b^2)$$

Similarly,

$$c_{f_{min}}{}^2 = \tfrac{1}{2}(a^2 + b^2 + |a^2 - b^2|)$$
$$= \text{Max}\,(a^2, b^2)$$

Thus, $c_f{}^2$ and $c_s{}^2$ are arranged as follows:

$$0 \le c_s{}^2 \le \text{Min}\,(a^2, b^2) \le \text{Max}\,(a^2, b^2) \le c_f{}^2 \le a^2 + b^2 \qquad (2.51)$$

or dividing by $b_x{}^2$,

$$0 \le M_s{}^2 \le \text{Min}\,(M_a{}^2,\ 1 + M_y{}^2) \le \text{Max}\,(M_a{}^2,\ 1 + M_y{}^2)$$
$$\le M_f{}^2 \le M_a{}^2 + M_y{}^2 + 1 \qquad (2.52)$$

The Limit of Strong Shocks.* For compressive shocks $1 \le \beta \le \beta_{max}$ if the subscript 2 in $\beta = \rho_2/\rho_1$ is taken to refer to the downstream side of the shock. The upper limit $\beta_{max}$ can be seen clearly from Equation 2.43 by noting that as the expression $[\beta + 1 - \gamma(\beta - 1)]$ in the denominator approaches zero, one of the roots of this bicubic must approach infinity if the equation is to remain in balance. Equating this expression to zero yields

$$\beta_{max} = \beta_{min} = \frac{\gamma + 1}{\gamma - 1} \qquad (2.53)$$

the familiar limit attained by the density ratio for ordinary shocks. It is interesting to note from Equation 2.36 that, as $M_{A_1} \to \infty$, $B_{y_2}/B_{y_1}$ attains the same limit as the density ratio.

---

* This limit corresponds to a strong slow shock of the type discussed by Shercliff.[14] The other strong shocks which he discusses are obtained from Equation 2.58 by setting $M_{s_1} = 0$.

It is of interest to explore the roots of Equation 2.43 for strong shocks in more detail. First, if we assume ${M_{A_1}}^2$ to be large and we drop comparative terms which remain bounded in the limit, Equation 2.43 becomes

$$ {M_{A_1}}^4 \left\{ {M_{A_1}}^2 \quad \frac{2\beta {M_{a_1}}^2 + \beta[2\beta - \gamma(\beta - 1)]{M_{y_1}}^2}{\beta + 1 - \gamma(\beta - 1)} \right\} = 0 \tag{2.54} $$

which gives an equation for computation of the velocity of strong shocks in terms of upstream state parameters and the density ratio.

Second, multiply Equation 2.43 by $[\beta + 1 - \gamma(\beta - 1)]$, and then let this factor vanish. The result is the following quadratic equation in ${M_{A_1}}^2$:

$$ {M_{a_1}}^2 \left( {M_{A_1}}^2 - \frac{\gamma + 1}{\gamma - 1} \right)^2 \quad \frac{{M_{y_1}}^2}{\gamma - 1} {M_{A_1}}^4 = 0 \tag{2.55} $$

which has the discriminant

$$ D \equiv 4{M_{a_1}}^4 \left( \frac{\gamma + 1}{\gamma - 1} \right)^2 - 4{M_{a_1}}^2 \left( {M_{a_1}}^2 + \frac{{M_{y_1}}^2}{\gamma - 1} \right) \left( \frac{\gamma + 1}{\gamma - 1} \right)^2 $$

$$ = -4{M_{a_1}}^2 {M_{y_1}}^2 \frac{(\gamma + 1)^2}{(\gamma - 1)^2} < 0 $$

Inasmuch as D is always negative ($\gamma > 1$), two of the roots of Equation 2.43 must have become complex before $\beta$ attains the limit $\beta_m$, indicating the nonexistence of intermediate shocks of great strength. We shall see later that the root which remains real can correspond to either a slow or a fast shock.

It would be interesting fo find $\beta_m$ for slow and intermediate shocks. The condition for this is that Equation 2.43 must have a double root, since for an infinitesimally higher value $\beta$, two roots of the equation become imaginary. The condition on a cubic equation for a double root is that its discriminant must be zero. If the cibic is represented in the form

$$ x^3 + bx^2 + cx + d = 0 \tag{2.56} $$

the discriminant is

$$ D \equiv 18bcd - 4b^3d + b^2c^2 - 4c^3 - 27d^2 \tag{2.57} $$

The coefficients b, c, d, easily obtained by comparing Equations 2.43 and 2.56 with $x = M_{A_1}^2$, are functions of $\beta$, $\gamma$, $M_{a_1}^2$, and $M_{y_1}^2$. Unfortunately, the equation $D = 0$ yields an extremely complicated relation among these variables, and can perhaps be solved numerically only in particular cases.

The Shock Adiabatic in the $M_{A_1}$-$M_{A_2}$ Plane. If $\beta$ is eliminated from Equation 2.43 by means of Equation 2.42, the result can be solved explicitly for $M_{A_1}^2$. Making use of Equation 2.48, the result is

$$M_{A_1}^2 = \frac{\left(M_{A_2}^2 - 1\right)^2\left[(\gamma+1)M_{A_2}^2 - 2M_{f_1}^2 M_{s_1}^2\right] - \left(M_{f_1}^2 - 1\right)\left(1 - M_{s_1}^2\right)\left(\gamma M_{A_2}^2 - \gamma - 1\right)M_{A_2}^2}{(\gamma-1)\left(M_{A_2}^2 - 1\right)^2 + \left(M_{f_1}^2 - 1\right)\left(1 - M_{s_1}^2\right)\left[(2-\gamma)M_{A_2}^2 + \gamma - 1\right]} \tag{2.58}$$

Thus, for a given $\gamma$, specification of the normalized speeds of fast and slow sound ahead of the shock results in a single curve along which all shocks must lie; fortunately, this curve can be found by straightforward computation without solving a cubic.

To complete the picture it is necessary also to know the speeds of fast and slow sound behind a given shock. To obtain them, first compute $M_{a_2}^2$ from either Equation 2.40 or 2.41. After substituting for $\beta$ from Equation 2.42 and using 2.48, the slightly simpler Equation 2.40 becomes

$$M_{a_2}^2 = M_{f_1}^2 M_{s_1}^2 + \gamma\left(M_{A_1}^2 - M_{A_2}^2\right)\left[1 - \frac{1}{2}\left(M_{f_1}^2 - 1\right)\left(1 - M_{s_1}^2\right)\frac{\left(M_{A_1}^2 + M_{A_2}^2 - 2\right)}{\left(M_{A_2}^2 - 1\right)^2}\right] \tag{2.59}$$

The quantity $M_{y_2}^2$ can be obtained by dividing both sides of Equation 2.36 by $B_x$. Thus,

$$M_{y_2}^2 = \left(\frac{M_{A_1}^2 - 1}{M_{A_2}^2 - 1}\right)^2 (M_{f_1}^2 - 1)(1 - M_{s_1}^2) \tag{2.60}$$

Then, $M_{f_2}^2$ and $M_{s_2}^2$ are obtained by substituting Equations 2.59 and 2.60 into Equation 2.49.

For calculation and manipulation of these quantities it is convenient to introduce the following notation:

$$\xi = M_{A_1}^2 - 1 \qquad \eta = M_{A_2}^2 - 1$$

$$f = M_{f_1}^{\ 2} \qquad\qquad s = M_{s_1}^{\ 2}$$

Making these substitutions, Equation 2.58 can take the form

$$\xi = \eta \left( \frac{\beta_m \eta^2 + p\eta - q}{\eta^2 + r\eta + q} \right) \tag{2.61}$$

in which

$$\left.\begin{aligned} \beta_m &= \frac{\gamma + 1}{\gamma - 1} \\ p &= r - \frac{2}{\gamma - 1}(f + s - 2) \\ q &= \frac{1}{\gamma - 1}(f - 1)(1 - s) \\ r &= \left(\frac{2 - \gamma}{\gamma + 1}\right)(f - 1)(1 - s) \end{aligned}\right\} \tag{2.62}$$

Equation 2.59 becomes

$$M_{a_2}^{\ 2} = fs + \gamma(\xi - \eta) \left[ 1 - \frac{1}{2}(f - 1)(1 - s)\frac{(\xi + \eta)}{\eta^2} \right] \tag{2.63}$$

and Equation 2.60 becomes

$$M_{y_2}^{\ 2} = (f - 1)(1 - s)\left(\frac{\xi}{\eta}\right)^2 \tag{2.64}$$

In Figure 2.10, the quantities $M_{A_2}$, $M_{f_2}$, and $M_{s_2}$ are plotted as functions of $M_{A_1}$ for a monatomic gas ($\gamma = 5/3$) and for a particular choice of the upstream state given by $f = 3/2$ and $s = 1/2$. Thus, it is assumed that index 1 refers to the upstream state. (For this case, the ratio of the gas pressure to magnetic pressure ahead of the shock is 0.72, and the magnetic field ahead of the shock is inclined at an angle of 26.5 degrees from the normal to the shock.) Equation 2.42 shows that lines of constant density ratio are rays from the origin in the $M_{A_1}$-$M_{A_2}$ plane. Thus, the ray $\beta = 1$ intersects the shock profile at the three small-disturbance speeds, the ray $\beta = 1.5$ is the maximum density ratio for slow and intermediate shocks,* and the ray

---

* The value $\beta = 1.5$ for the point of tangency is not exact, and it is a coincidence that in this case $f = 1.5$ also.

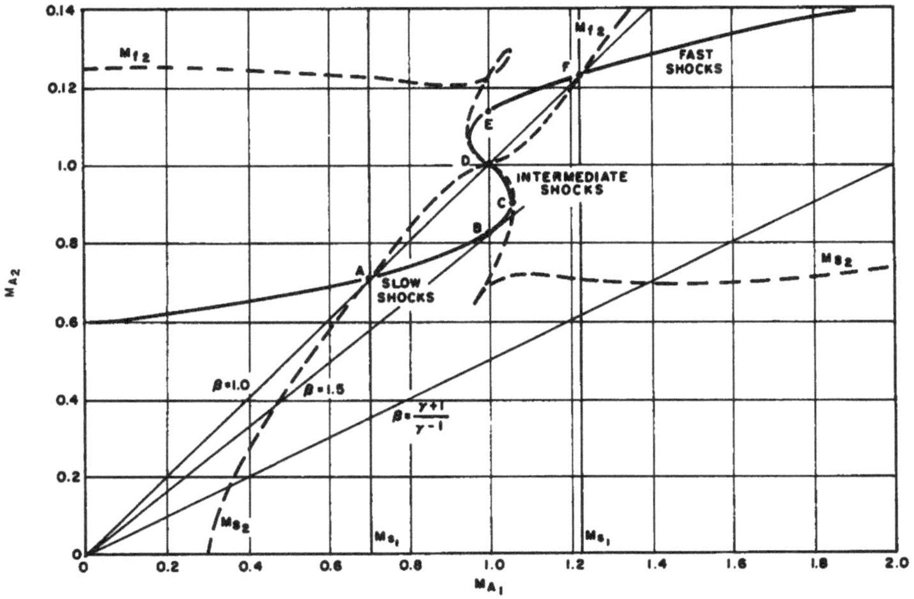


Figure 2.10. The shock-adiabatic curve for a particular choice of $M_{f_1}$ and $M_{s_1}$

$\beta = (\gamma + 1)/(\gamma - 1)$ is the asymptote for the fast-shock curve. The parts of the shock curve above the line $\beta = 1$ correspond to expansion shocks and have been shown to violate the second law of thermodynamics. In accord with the definitions given on page 13, the region of the shock-adiabatic curve of Figure 2.10 between A and B is the locus of slow shocks, the region BCD is the locus of intermediate shocks, and the region from point F to infinity is the locus of fast shocks.

For each shock, given by a point $(M_{A_1}, M_{A_2})$, the fast and slow sound speeds behind the shock are obtained by following the ordinate up or down to the two dashed curves labeled $M_{f_2}$ and $M_{s_2}$. Note that the previously derived relation $M_{s_2} \leq 1 \leq M_{f_2}$ holds true. For very strong fast shocks, the asymptotic values of $M_{f_2}$ and $M_{s_2}$ are obtained from Equations 2.61, 2.63, 2.64, and 2.49 by assuming that $\eta >> 1$. In this case, Equation 2.61 shows that $\xi \to \beta_m \eta$. Then Equation 2.63 reduces to

$$M_{a_2}^{\ 2} \to \gamma(\beta_m - 1)\eta = \frac{2\gamma}{\gamma - 1}\eta \to \frac{2\gamma}{\gamma - 1} M_{A_2}^{\ 2}$$

and Equation 2.64 shows that $M_{y_2}^{\ 2}$ stays finite. Using this information, one can easily see from Equation 2.49 that

$$M_{f_2}{}^2 \to M_{a_2}{}^2$$

and

$$M_{s_2}{}^2 \to 1$$

as the shock velocity goes to infinity.

When the indices 1 and 2 correspond to the states ahead of and behind the shock, respectively, as was stated in the above discussion, all shocks below the line $\beta = 1$ are compressive, and all shocks above are expansive. These latter shocks therefore cannot exist in nature because they violate the second law of thermodynamics. There is no reason, however, why the indices of Figure 2.10 cannot be given opposite meanings; i.e., let index 1 correspond to the downstream state and index 2 to the upstream state. Then Figure 2.10 gives all the shock adiabatics for which the downstream state is fixed. In this case all shocks below the line $\beta = 1$ are expansive, and those above are compressive. Thus, the branch of the adiabatic from the far left to point A becomes the locus of slow shocks, and the branch from E to F the locus of fast shocks. The locus from D to E becomes the locus of compressive intermediate shocks. The slow-shock branch in this context is most interesting because there appears to be no upper limit on the density ratio. There is a practical limit, however, because the $M_{s_2}$ curve goes to zero, corresponding — from Equation 2.48 — to zero sound speed in the fluid ahead of the shock, and hence to zero temperature and pressure.* In the broader interpretation of Figure 2.10 just described, the "switch-on" and "switch-off" shocks, defined earlier, are clearly the points E and B, respectively.

The complex behavior of the curves of Figure 2.10 can be seen in relation to the discussion on page 13 by distorting them topologically so that $M_{f_2}$ and $M_{s_2}$ appear as horizontal lines. The result is shown schematically in Figure 2.11. Part of this diagram is given by Polovin.[17] The numbers and arrows indicate transitions, discussed earlier, between the points shown in Figures 2.5 and 2.6. An interesting feature of Figure 2.11 is that in this case 1 → 4 shocks cannot be stable because they do not satisfy the conservation laws; shocks which do satisfy these laws fall on the indicated curves. It is important for the work of Chapter 5 to determine whether or not this is a general property, and we shall now show that it is not.

---

* This limit corresponds to a strong slow shock of the type discussed by Shercliff.[14] The other strong shocks which he discusses are obtained from Equation 2.58 by setting $M_{s_1} = 0$.

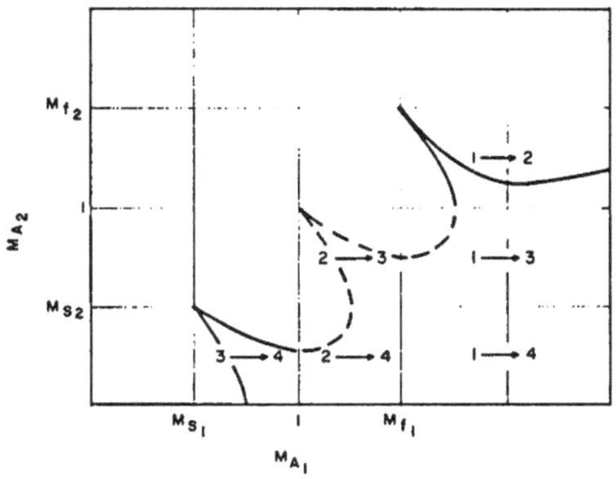


Figure 2.11. Schematic representation of the relation of the shock curve to the velocities of small disturbances

A study of Figure 2.10 will show that the extremum point C of the shock adiabatic must always coincide with the line $M_{A_2} = M_{s_2}$. Therefore, an easy way to determine if the shock adiabatic can lie partially in the 1 → 4 region is to determine if there can be three real solutions to the shock-adiabatic equation (2.61) along the line $M_{A_1} = M_{f_1}$ ($\xi = f - 1$). To simplify the algebra, define

$$\left.\begin{aligned} \mu &= f - 1 \\ \nu &= 1 - s \end{aligned}\right\} \qquad (2.65)$$

and note that

$$\left.\begin{aligned} 1 &< \mu < \infty \\ 0 &< \nu < 1 \end{aligned}\right\} \qquad (2.66)$$

Substituting $\xi = \mu$, we can write Equation 2.61 in the form

$$\beta_m \eta^3 + (p - \mu)\eta^2 - (q + \mu r)\eta - \mu q = 0 \qquad (2.67)$$

but since one root of this equation is $\eta = \mu$, it must be factorable. Making use of Definitions 2.62, it is easy to show that the factors can be written

$$(\eta - \mu)\{(\gamma + 1)\eta^2 + \nu[(2 - \gamma)\mu + 2]\eta + \mu\nu\} = 0 \qquad (2.68)$$

If the discriminant of the quadratic factor can be real, 1 → 4 shocks are possible. The discriminant is

$$D = \nu^2[(2 - \gamma)\mu + 2]^2 - 4(\gamma + 1)\mu\nu$$

It can be either positive or negative, and the bordering case when point C in Figure 2.10 just touches the line $\xi = \mu$ is obtained by setting D = 0. This gives the following relation between $\mu$ and $\nu$:

$$\nu' = \frac{4\mu'}{(\mu' + 2)^2} \qquad (2.69)$$

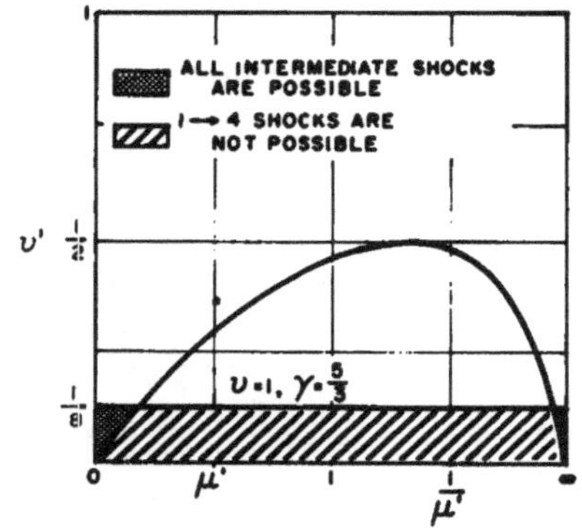


Figure 2. 12. Regions in which the 1 → 4 type intermediate shocks are possible

in which

$$\nu' = \left(\frac{2 - \gamma}{\gamma + 1}\right)\nu, \quad \mu' = (2 - \gamma)\mu$$

Equation 2. 69 is plotted in Figure 2. 12, in which 1 → 4 shocks are possible only in the dotted regions. Note that at the maximum value of $\nu (\nu = 1)$, $0 < \nu' < 1/2$ since $1 < \gamma < 2$. The horizontal line $\nu' = 1/8$ is for $\nu = 1$, $\gamma = 5/3$.

## Chapter 3

# STABILITY OF SHOCKS WITH RESPECT TO SMALL DISTURBANCES

In this chapter, a shock wave is still treated as a mathematical discontinuity in a nondissipative, perfectly conducting fluid. An idealized experiment is presented in which a plane monochromatic wave of infinitesimal amplitude strikes a plane shock wave of infinite extent; then the conditions under which the subsequent disturbances remain small are taken as necessary conditions for the stability of the shock. There is no loss of generality in specializing the incident wave in this way, as any arbitrary small disturbance is a sum of plane waves.

Before looking into the details of this problem for oblique magnetohydrodynamic shocks, it is worth while to discuss the corresponding problem in ordinary hydrodynamics. To fix the ideas, let us assume that a plane monochromatic sound wave is normally incident on the upstream side of the shock. In general, when this incident wave strikes the shock, both reflected and refracted waves will be produced. To each divergent wave there corresponds a single unknown constant to be determined in relation to the amplitude of the incident wave from the boundary conditions at the perturbed shock. There are three boundary conditions, obtained from the conservation of mass, momentum, and energy; but since they include the perturbed shock velocity as an unknown, only two equations remain from which to determine the amplitudes of the divergent waves. One of these waves will always be an entropy wave carried downstream with the fluid; thus, there is only one condition left from which to determine a divergent sound wave amplitude.

As is easily verified by comparing the velocity of sound with the velocity of the flow with respect to the shock, there is exactly one divergent sound wave only in the case where the flow is supersonic upstream of the shock and subsonic downstream, i. e., the case which satisfies the conservation laws in the steady state, and the requirement that entropy increase. For all other choices of the upstream and downstream parameters there is more than one divergent sound wave, so that not one but an infinite number of solutions for the divergent waves are possible.

Looking at the problem in another way, if the amplitude of the incident wave is zero and there are more divergent waves than boundary conditions, nonzero values for the amplitudes of the divergent waves can still be computed. Thus, the condition for nonuniqueness of the solution for divergent waves is also the condition for spontaneous emission of sound waves from the shock, and hence a sufficient condition for instability, because such emission must necessarily reduce the shock strength to zero. This is not surprising in ordinary hydrodynamics, since the conditions that the conservation laws are not satisfied and/or the entropy decreases across the shock are in themselves sufficient conditions for instability.

Now, if the flow velocity is supersonic upstream and subsonic downstream, the amplitudes of the two divergent waves can be calculated, and since no dissipative mechanisms have been assumed outside the shock, the perturbations are proportional to the amplitude of the incident wave, as no complex factors are present. The perturbations therefore die out when the incident wave dies out, and with the framework of the assumptions made, stability is predicted. The problem just described was solved by Burgers[12] and is discussed by Landau and Lifshitz (on pages 328-329 of their book.[18])

The problem of stability of ordinary shocks was considered from a different point of view by N. E. Kotchine, of Leningrad.[19] Kotchine looked for unique solutions to the steady-flow conservation equations (Rankine-Hugoniot equations) when the upstream and downstream conditions were arbitrary; and when just one discontinuity was present, he found them only when the flow was supersonic upstream and subsonic downstream of the discontinuity. When the reverse was true, for example, he found a unique solution if he postulated three discontinuities: a shock moving upstream, an expansion wave moving downstream, and a stationary discontinuity in between. He also considered initial discontinuities which were either supersonic or subsonic on both sides, and found other combinations of shocks and expansion waves which satisfied them, thus providing more insight into the nature of the instability. Specifically, he showed that the initial unstable shock wave breaks up into a sequence of stable shock waves. Note that an essential point in this type of analysis is that the several waves or discontinuities which uniquely satisfy the initial upstream and downstream conditions must in themselves be stable. This is true for all but the stationary or contact discontinuity, across which there is no flow or pressure jump, only a density jump. Since there is nothing to counteract the diffusion of gas from the high- to the low-density side, the contact discontinuity has no steady-state structure, and hence cannot be considered stable. This will cast some doubt on the validity of the type of analysis performed by Kotchine and recently by Gogosov[20,21,22] for magnetohydrodynamic shocks until the effect of diffusion of the stationary discontinuity is understood.

The type of unstable shock described here has been appropriately called "nonevolutionary" in the recent Russian literature.[17] Thus, a nonevolutionary shock is one which is not only unstable but for which it is impossible to calculate the growth rate of the perturbations even an instant after they arise, because there is no unique solution to the small-disturbance problem. An "evolutionary" shock may be unstable, but at least it is possible to compute its growth characteristics from an infinitesimal disturbance.

For the stability of magnetohydrodynamic shock waves, to be discussed in detail in this chapter, the same general considerations arise, except that the situation is much more complicated owing chiefly to the existence of seven types of waves instead of three. Also, a remarkable aspect appears in that there are nonevolutionary situations for which there are not too many but too few waves to satisfy the boundary conditions at the shock. Thus, the amplitudes of the divergent waves could be computed only if one or more of the boundary conditions were ignored. Furthermore, in this case, spontaneous emission of waves is not possible. This and other aspects of the problem of stability of magnetohydrodynamic shocks in perfect fluids will be discussed in this chapter.

A few more remarks related to the meaning of the term "stability" as used in this chapter are appropriate. A shock is said to be stable if, in the presence of arbitrary small excitations, it retains its original form as a single surface of discontinuity, and if all of the flow disturbances remain small. The term "unstable" is always used in the stronger sense of "nonevolutionary." In the presence of small disturbances, a nonevolutionary shock wave cannot retain its original form; it must disintegrate either into a series of stable discontinuities or into a more general unsteady-flow pattern.

To lay a proper foundation, we shall first present briefly the equations of perfect-fluid magnetohydrodynamics, then give the linearized form for small disturbances, and derive the boundary conditions for three-dimensional disturbances at the shock. Because of their importance in Section 4, we shall discuss concepts of phase and group velocity and give an explicit proof of the fact that the group velocity is identical with the energy-propagation velocity for magnetoacoustic and Alfvén waves. Then we explain in detail the influence of small normal disturbances on magnetohydrodynamic shocks and follow this with the proof of Kontorovich[23] that the results apply without modification to the case of arbitrary three-dimensional disturbances. Finally, we shall examine work of Liubarskii and Polovin[24] and Gogosov[20,21,22] on the disintegration of nonevolutionary shocks.

## 1. Small-Perturbation Equations and Boundary Conditions

We give the equations of magnetohydrodynamics here in sufficient generality for this chapter only; that is, we assume that the fluid is a nonviscous continuum, a nonconductor of heat, and a perfect conductor of electricity. Moreover, we assume that the local flow velocity is nonrelativistic. Derivations of these equations may be found in many sources, for example in Reference 1. For convenience, we use tensor notation throughout, even representing components of cross products in the form

$$(\vec{A} \times \vec{B})_i = \epsilon_{ijk} A_j B_k$$

where $\epsilon_{ijk}$ are the permutation symbols defined so that the component $\epsilon_{123} = 1$, and $\epsilon_{ijk}$ is also unity for every even number of permutations of the indices from this basic position. For an odd number of permutations it is minus one, and when any two or all three of the indices are equal, it is zero. Use of the three-index permutation symbols is facilitated because of the identity[25]

$$\epsilon_{ijk}\epsilon_{i\ell m} = \delta_{j\ell}\delta_{km} - \delta_{jm}\delta_{k\ell} \tag{3.1}$$

which is used freely as occasions arise, and in which $\delta_{ij}$ is the ordinary Kronecker delta.

Basic Equations. The equations of perfect-fluid magnetohydrodynamics are expressions of the continuity of mass, momentum, and energy at any point in the flow, coupled with the macroscopic form of Maxwell's equations for the electromagnetic field due to distributed charges and currents. Inasmuch as we are considering a range of plasmas in which the effects of free currents are of much greater importance than magnetization currents, we shall assume that the magnetic permeability $\mu$ everywhere has its free-space value $4\pi(10)^{-7}$.

When the flow velocity is nonrelativistic, the charge separation outside of shocks is very small, and consequently the displacement current is small (of the order $u^2/c^2$) compared with other terms. Also, in unsteady-flow problems, the electric part of the Lorentz force and the electrical field energy are of the order $u^2/c^2$ compared with their magnetic counterparts.

When terms of the order $u^2/c^2$ are neglected, the basic equations of perfect-fluid magnetohydrodynamics are

continuity:
$$\frac{\partial \rho}{\partial t} + \frac{\partial \rho v_k}{\partial x_k} = 0 \tag{3.2}$$

momentum:
$$\rho \frac{\partial v_i}{\partial t} + \rho v_k \frac{\partial v_i}{\partial x_k} + \frac{\partial p}{\partial x_i} = \epsilon_{ik\ell} j_k B_\ell \tag{3.3}$$

energy:* $$\frac{\partial}{\partial t}\left[\rho\left(U + \frac{v^2}{2}\right) + \frac{B^2}{2\mu}\right] + \frac{\partial}{\partial x_k}\left[\rho v_k\left(h + \frac{v^2}{2}\right) + \epsilon_{k\ell m}\frac{E_\ell B_m}{\mu}\right] = 0 \tag{3.4}$$

magnetic field: $$\epsilon_{ijk}\frac{\partial B_k}{\partial x_j} = \mu j_i \tag{3.5}$$

Faraday's law of induction: $$\epsilon_{ijk}\frac{\partial E_k}{\partial x_j} + \frac{\partial B_i}{\partial t} = 0 \tag{3.6}$$

Since we assume infinite conductivity, the requirement of finite currents presupposes that there can be no electrical field in a coordinate system at rest with respect to the fluid. This means that in laboratory coordinates,

$$E_i + \epsilon_{ijk} v_j B_k = 0 \tag{3.7}$$

Substituting Equation 3.7 in 3.4 and 3.6, Equation 3.5 in 3.3, and using Equation 3.2 in 3.3, one obtains the basic equations

$$\frac{\partial \rho}{\partial t} + \frac{\partial \rho v_k}{\partial x_k} = 0 \tag{3.2}$$

$$\frac{\partial \rho v_i}{\partial t} + \frac{\partial}{\partial x_k}\left[\rho v_i v_k + \left(p + \frac{B^2}{2\mu}\right)\delta_{ik} - \frac{B_i B_k}{\mu}\right] = 0 \tag{3.3a}$$

$$\frac{\partial}{\partial t}\left[\rho\left(U + \frac{v^2}{2}\right) + \frac{B^2}{2\mu}\right] + \frac{\partial}{\partial x_k}\left[\rho v_k\left(h + \frac{v^2}{2}\right) + \frac{B^2}{\mu} v_k \quad \frac{B_j v_j}{\mu} B_k\right] = 0 \tag{3.4a}$$

$$\frac{\partial B_i}{\partial t} + \frac{\partial}{\partial x_k}\left[v_k B_i - v_i B_k\right] = 0 \tag{3.8}$$

Taking the divergence of Equation 3.8, one can see that $\partial B_i/\partial x_i$ is a function of space variables only. The experimental fact that magnetic charges do not exist in nature then is expressed as

---

* The energy equation can be obtained by adding the ordinary hydrodynamic energy equation and the Poynting equation; however, a rigorous proof of this is not easy. There is an excellent discussion of this by Boa-Teh Chu.[26]

$$\frac{\partial B_i}{\partial x_i} = 0 \tag{3.9}$$

Equation 3.9 is not an extra equation but an alternate form for one of the three equations 3.8.

With the help of Equations 3.2, 3.3, 3.8, and 3.9, Equation 3.4a can be manipulated into the form

$$\frac{DU}{Dt} + p\frac{D}{Dt}\left(\frac{1}{\rho}\right) = T\frac{Ds}{Dt} = 0$$

in which

$$\frac{D}{Dt} = \frac{\partial}{\partial t} + v_k\frac{\partial}{\partial x_k}$$

is the substantial derivative. Thus, it is often useful to replace Equation 3.4 by

$$\frac{\partial s}{\partial t} + v_k\frac{\partial s}{\partial x_k} = 0 \tag{3.10}$$

which shows that because dissipation has been neglected, the entropy of a given mass of fluid remains constant. Equation 3.10, of course, does not hold true across a shock, whereas Equation 3.4 does.

Equations for Small Perturbations in the Flow. As a preparation for the analysis of shock stability, linear equations are given in this paragraph for small perturbations from a state in which all variables are functions of neither space nor time. Each variable is written in the form $\rho = \rho_0 + \rho'$, where the zero denotes the steady-state quantity and the prime denotes the small perturbation; however, since there is no reason for confusion, the subscript zero is dropped.

If each dependent variable, expressed in the above form, is substituted into Equations 3.2, 3.3a, 3.8, 3.10, and 3.9, and all second and higher terms are dropped, the results can be written as follows:

$$\frac{\partial \rho'}{\partial t} + v_k\frac{\partial \rho'}{\partial x_k} + \rho\frac{\partial v_k'}{\partial x_k} = 0 \tag{3.11}$$

$$\rho\frac{\partial v_i'}{\partial t} + \rho v_k\frac{\partial v_i'}{\partial x_k} + a^2\frac{\partial \rho'}{\partial x_i} + p_s\frac{\partial s'}{\partial x_i} + \frac{B_k}{\mu}\frac{\partial B_k'}{\partial x_i} - \frac{B_k}{\mu}\frac{\partial B_i'}{\partial x_k} = 0 \tag{3.12}$$

$$\frac{\partial B_i'}{\partial t} + v_k \frac{\partial B_i'}{\partial x_k} + B_i \frac{\partial v_k'}{\partial x_k} - B_k \frac{\partial v_i'}{\partial x_k} = 0 \tag{3.13}$$

$$\frac{\partial s'}{\partial t} + v_k \frac{\partial s'}{\partial x_k} = 0 \tag{3.14}$$

$$\frac{\partial B_i'}{\partial x_i} = 0 \tag{3.15}$$

in which

$$a^2 = \left(\frac{\partial p}{\partial \rho}\right)_s , \qquad p_s = \left(\frac{\partial p}{\partial s}\right)_\rho \tag{3.16}$$

Boundary Conditions at the Shock Wave. One of the procedures used in studying shock stability is to allow a small-disturbance wave to strike the shock, find which waves can be reflected and refracted and the directions in which they go, and then if possible determine the amplitudes of these waves. For the latter problem it is necessary to have the boundary conditions at the shock in detailed form. The method used here to obtain them has aspects similar to the ones used by Burgers[12] to derive the boundary conditions for normal disturbances falling on ordinary shocks.

When a plane wave strikes a shock, momentum can be conserved only if the vectors representing the directions of propagation of the incident, reflected, and refracted waves all lie in a single plane perpendicular to the plane of the shock. This fact can also be seen in an alternate way by noting that the propagation vectors* of all of the waves must have identical projections on the plane of the shock. For convenience in this derivation one may take the plane of the propagation vectors as the x-y plane and assume that the undisturbed shock lies in the y-z plane. As shown in Figure 3.1, there are disturbances therefore only in the x- and y-directions. The undisturbed magnetic field and flow velocity, however, generally have components in all three directions.

To obtain the boundary conditions, consider a portion of the shock of infinitesimal width ($\delta$) in the y-direction and unit length in the z-direction. The strip is located at $y = y_0$, as shown in Figure 3.2. The detailed motion of the portion $\delta$ as it moves back and forth under the influence of a small disturbance is

---

* The propagation vector $\vec{k}$ has a magnitude equal to the reciprocal wave length and points in the direction of travel of the wave.

followed by means of two new reference frames. The first is translational,

$$\left.\begin{aligned} x' &= x \quad \zeta(y_0, t) \\ y' &= y - y_0 \\ z' &= z \\ t' &= t \end{aligned}\right\} \qquad (3.17)$$

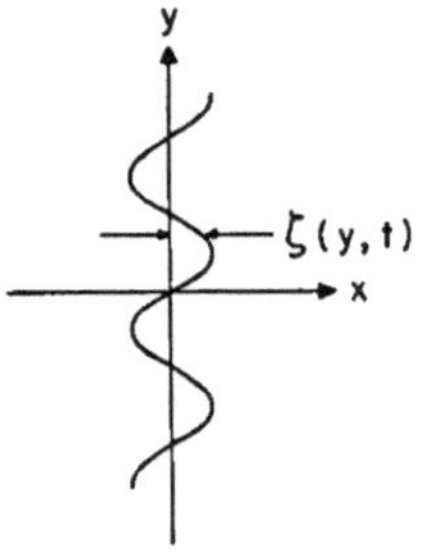


Figure 3.1. The disturbed shock

and the second is rotational,

$$\left.\begin{aligned} x'' &= x' - \theta y' \\ y'' &= \theta x' + y' \\ z'' &= z' \\ t'' &= t' \end{aligned}\right\} \qquad (3.18)$$

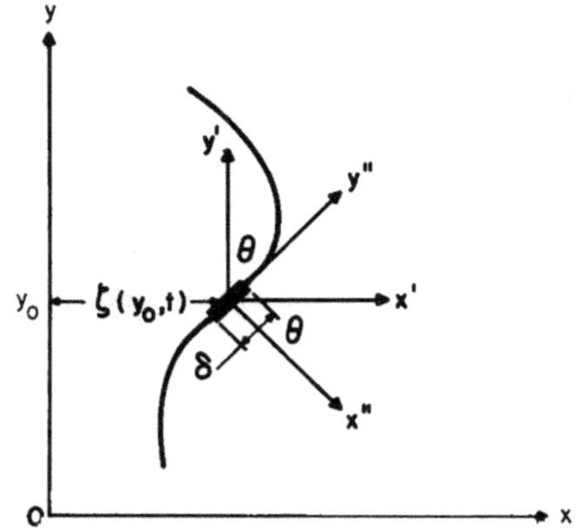


Figure 3.2. An infinitesimal portion of the disturbed shock

in which, in keeping with small-disturbance theory, it is tacitly assumed that $\theta$ is a small angle, given from Figure 3.2 by

$$\theta = \frac{\partial \zeta}{\partial y}(y_0, t) \qquad (3.19)$$

The independent variables $x, y, z, t$ in the equations of motion are now replaced by the new variables $x''$, $y''$, $z''$, $t''$. We emphasize that the vectors $\vec{v}$ and $\vec{B}$ are not transformed to the new coordinates; they remain referred to the $x, y, z, t$ reference frame, and their components are merely expressed in terms of the new coordinates. The derivative operators in the equation of motion are therefore replaced by the following expressions: From Equation 3.17,

$$\frac{\partial}{\partial x} = \frac{\partial}{\partial x'}, \qquad \frac{\partial}{\partial y} = \frac{\partial}{\partial y'}$$

$$\frac{\partial}{\partial t} = \frac{\partial}{\partial t'} - \frac{\partial \zeta}{\partial t}\frac{\partial}{\partial x'}$$

and from Equation 3. 18,

$$\frac{\partial}{\partial x'} = \frac{\partial}{\partial x''} + \frac{\partial \zeta}{\partial y}\frac{\partial}{\partial y''}$$

$$\frac{\partial}{\partial y'} = \frac{\partial}{\partial y''} - \frac{\partial \zeta}{\partial y}\frac{\partial}{\partial x''}$$

$$\frac{\partial}{\partial t'} = \frac{\partial}{\partial t''} - y'\frac{\partial \theta}{\partial t'}\frac{\partial}{\partial x''} + x'\frac{\partial \theta}{\partial t'}\frac{\partial}{\partial y''}$$

Combining, we have

$$\left.\begin{aligned}
\frac{\partial}{\partial x} &= \frac{\partial}{\partial x''} + \frac{\partial \zeta}{\partial y}\frac{\partial}{\partial y''}\\
\frac{\partial}{\partial y} &= \frac{\partial}{\partial y''} - \frac{\partial \zeta}{\partial y}\frac{\partial}{\partial x''}\\
\frac{\partial}{\partial t} &= \frac{\partial}{\partial t''} - \frac{\partial \zeta}{\partial t}\frac{\partial}{\partial x''} - \frac{\partial \zeta}{\partial t}\frac{\partial \zeta}{\partial y}\frac{\partial}{\partial y''} - y'\frac{\partial \theta}{\partial t'}\frac{\partial}{\partial x''} + x'\frac{\partial \theta}{\partial t'}\frac{\partial}{\partial y''}
\end{aligned}\right\} \tag{3.20}$$

After substituting Equation 3. 20 into Equations 3. 2, 3. 3a, 3. 4a, 3. 8, and 3. 9, we integrate these equations throughout a small volume defined by

$$-\epsilon \leq x'' \leq \epsilon$$

$$-\delta/2 \leq y'' \leq \delta/2$$

$$0 \leq z'' \leq 1$$

in which $\epsilon$ will later be made to vanish. But in the new coordinates, the dependent variables $\rho$, $v_i$, and $B_i$ in the above volume vary rapidly with $x''$ and $t''$ but slowly with $y''$ and $t''$ Hence, integrations like

$$\int_{-\epsilon}^{\epsilon} \frac{\partial \rho}{\partial y''}\, dx'' \qquad \text{and} \qquad \int_{-\epsilon}^{\epsilon} \frac{\partial \rho}{\partial t''}\, dx''$$

vanish as $\epsilon$ vanishes, but integrations like

$$\int_{-\epsilon}^{\epsilon} \frac{\partial \rho}{\partial x''}\, dx'' = \rho(\epsilon) - \rho(-\epsilon) = \{\rho\}$$

remain finite because of the discontinuity. Since the terms $\partial\zeta/\partial y$ and $\partial\zeta/\partial t$ in Equation 3.20 depend on $x''$ only to the second order, they can be removed from under the integral sign.

From these expressions, it is clear that the desired boundary conditions can be obtained from Equations 3.2, 3.3a, 3.4a, 3.8, and 3.9 by making the substitutions

$$\left.\begin{aligned} \frac{\partial}{\partial x} &\rightarrow \{\ \} \\ \frac{\partial}{\partial y} &\rightarrow -\frac{\partial\zeta}{\partial y}\{\ \} \\ \frac{\partial}{\partial z} &\rightarrow 0 \\ \frac{\partial}{\partial t} &\rightarrow -\frac{\partial\zeta}{\partial t}\{\ \} \end{aligned}\right\} \qquad (3.21)$$

in which the term proportional to $y'\ \partial\theta/\partial t'$ in the third part of Equation 3.20 is dropped because $\theta$ is a small angle and $y'$ is infinitesimal in the region over which the integration is carried out.

Before proceeding, note that Equations 3.4a and 3.8 are not valid through the shock, since the infinite-conductivity condition (in 3.7) has been used in deriving them from Equations 3.4 and 3.6, respectively. Within the braces { }, however, only values outside the shock layer are used; therefore it is perfectly valid to substitute the expressions in 3.21 into 3.4a and 3.8 to obtain the boundary conditions. Once the substitutions have been made, the independent variables $x''$, $y''$, $z''$, $t''$ are in principle replaced by $x$, $y$, $z$, $t$, so that the boundary conditions are expressed completely in terms of the original coordinates.

Now if we substitute Expressions 3.21 as indicated above and assume the perturbations in the resulting equations to be small, we obtain the desired boundary conditions. If we use the steady-state boundary conditions (Equations 2.1 through 2.6) to simplify them, we may manipulate them into the following form:

$$\{G_n'\} \equiv \{\rho(v_x' - \zeta_t) + v_x\rho' - \rho v_y\zeta_y\} = 0 \qquad (3.22)$$

$$\left\{v_x^2\rho' + 2Gv_x' + p^{*\prime} - \frac{B_x}{\mu}B_x' = 0\right\} \qquad (3.23)$$

$$\left\{G_n'v_y + Gv_y' - \frac{B_x}{\mu}B_y' - \frac{B_y}{\mu}B_n' - p^*\zeta_y\right\} = 0 \qquad (3.24)$$

$$\left\{ G_n' v_z + G v_z' - \frac{B_x}{\mu} B_z' - \frac{B_z}{\mu} B_n' \right\} = 0 \qquad (3.25)$$

$$\left\{ G_n' \overline{U}^* + G \overline{U}^{*\prime} + (p^* v_x)' - p^* v_y \zeta_y - \frac{B_x}{\mu} (\vec{B}_t \quad \vec{v}_t)' \quad \frac{(\vec{B}_t \cdot \vec{v}_t)}{\mu} B_n' \right\} = 0 \qquad (3.26)$$

$$B_y (v_x' \quad \zeta_t) + v_x B_y' - B_x v_y' - v_y B_x' = 0 \qquad (3.27)$$

$$B_z (v_x' - \zeta_t) + v_x B_z' - B_x v_z' - v_y B_z \zeta_y \quad v_z B_n' = 0 \qquad (3.28)$$

$$\{B_n'\} \equiv \{B_x' - B_y \zeta_y\} = 0 \qquad (3.29)$$

in which

$$\left.\begin{aligned}
G &= \rho v_x \\
p^* &= p + \frac{B_y^2 + B_z^2}{2\mu} \\
\overline{U}^* &= U + \frac{\vec{v}^2}{2} + \frac{B_y^2 + B_z^2}{2\mu\rho} \\
\vec{B}_t \cdot \vec{v}_t &= B_y v_y + B_z v_z \\
\zeta_t &= \frac{\partial \zeta}{\partial t} \\
\zeta_y &= \frac{\partial \zeta}{\partial y}
\end{aligned}\right\} \qquad (3.30)$$

Although a specific portion of the shock at $y = y_0$ was considered in the derivation, there is no mention of either $y_0$ or $\delta$ in the final equations; hence, they are valid everywhere. Note that the $x^{th}$ of Equations 3.8 gives no information at all and therefore does not appear. For this reason Equation 3.29, derived from 3.9, is no longer merely an alternate way of writing one of Equations 3.8 but must be included as one of the boundary conditions. Thus, there are eight boundary conditions, 3.22 through 3.29, containing ten perturbation quantities: two thermodynamic state variables, $v_i'$, $B_i'$, $\zeta_t$, and $\zeta_y$. As expected, these boundary conditions express the continuity of mass, momentum, energy, the tangential component of the electric field, and the normal component of the magnetic field. Elimination of $\zeta_t$ and $\zeta_y$ then leaves six equations from which to determine the amplitudes of the diverging waves.

Strictly speaking, the boundary conditions apply at $x = \zeta(y, t)$; however, since a typical perturbed quantity can be written at this point as

$$\rho'\Big]_{x=\zeta} = \rho'(0) + \frac{\partial \rho'}{\partial x}\zeta + \cdots$$

The error in applying the boundary conditions at $x = 0$ is of the order of terms already neglected.

Notice that in general the boundary conditions are completely coupled. If, however, the plane of the disturbance vectors coincides with the plane of the magnetic-field and velocity vectors (a possible condition), $v_z = B_z = 0$, and Equations 3.25 and 3.28 uncouple from the rest.

## 2. The Phase and Group Velocities of Small-Amplitude Waves

The concepts of phase and group velocity play a vital role in later developments in this chapter. As the properties of these concepts which are required in this chapter cannot easily be gleaned from the literature, a rather complete account of them is given here.

Eigenvalues and Eigenvectors. For the work of this section, the eigenvalues and eigenvectors of small-amplitude monochromatic plane waves are needed. Consider the formulation given in Equations 3.11 through 3.15. Assume that all perturbation quantities in those linear equations are proportional to

$$\exp\left[i(k_i x_i - \omega t)\right] \tag{3.31}$$

Then, if the notation $(\tilde{\rho}', \tilde{v}_i', \text{etc.})$ is used for the amplitudes of the perturbations, Equations 3.11 through 3.15 reduce to the following algebraic system:

$$-\omega_0 \tilde{\rho}' + \rho(k\tilde{v}') = 0 \tag{3.32}$$

$$-\rho\omega_0 \tilde{v}_i' + a^2 k_i \tilde{\rho}' + p_s k_i \tilde{s}' + k_i \frac{(B\tilde{B}')}{\mu} - \frac{(Bk)}{\mu}\tilde{B}_i' = 0 \tag{3.33}$$

$$-\omega_0 \tilde{B}_i' + B_i(k\tilde{v}') - (Bk)\tilde{v}_i' = 0 \tag{3.34}$$

$$-\omega_0 \tilde{s}' = 0 \tag{3.35}$$

$$(k\tilde{B}') = 0 \tag{3.36}$$

in which, following Syrovatskii,[27]

$$\omega_0 = \omega - (k\,v) \tag{3.37}$$

and the convenient notation $(k\,v) = k_i v_i$ has been used.

a. Entropy waves: Equation 3.35 allows two alternatives. If the entropy fluctuation $s' \neq 0$, then $\omega_0 = 0$, and the following results are obtained:

$$\left.\begin{aligned} &\tilde{s}' \neq 0 \\ &\tilde{v}_i' = \tilde{B}_i' = a^2\tilde{\rho}' + p_s\tilde{s}' \equiv \tilde{p}' = 0 \end{aligned}\right\} \tag{3.38}$$

Thus, fluctuations are possible in the thermodynamic variables such that the pressure remains constant. The eigenvector of the entropy wave, which has an eigenvalue $\omega_0 = 0$, is defined by Equation 3.38.

b. Alfvén waves: A solution of Equations 3.32-3.36 is sought in which there is no change in state variables, i.e., in which $\tilde{\rho}' = \tilde{s}' = 0$. Then from Equation 3.32, $(k\,\tilde{v}') = 0$, and if the inner product of Equation 3.33 with $k_i$ is taken, it is seen by using Equation 3.36 that $(B\,\tilde{B}') = 0$. Using these properties, Equations 3.33 and 3.34 become

$$\begin{bmatrix} \omega_0 & \dfrac{(B\,k)}{\mu\,\rho} \\ (B\,k) & \omega_0 \end{bmatrix} \begin{vmatrix} \tilde{v}_i' \\ \tilde{B}_i' \end{vmatrix} = 0 \tag{3.39}$$

Hence, a nontrivial solution with $\rho' = s' = 0$ is possible if

$$\omega_0 = \pm\,(b\,k) \tag{3.40}$$

in which

$$b_i = \frac{B_i}{\sqrt{\mu\,\rho}}$$

Since the frequency of a simple harmonic wave in a coordinate system at rest with respect to the fluid may be taken as a positive number with the direction of the wave given by $\vec{k}$, the upper sign in Equation 3.40 is taken when $(B\,k) > 0$, and vice versa. Substituting Equation 3.40 into one of 3.39, we obtain

$$\tilde{v}_i' = \mp\,\tilde{b}_i' \tag{3.41}$$

which, together with $\rho' = s' = 0$, gives the eigenvector of magnetohydrodynamic (or Alfvén) waves. Thus, the velocity and magnetic-field disturbances are 180 degrees out of phase when the wave propagates in the direction of $\vec{B}$, and they are in phase in the opposite case. The properties $(k\tilde{v}') = (B\tilde{B}') = 0$ show that the disturbances are transverse to both the direction of propagation and the direction of the magnetic field. All of these properties can be clearly visualized and confirmed by considering the B-lines as lines of tension in a fluid.

When the small-disturbance assumption is dropped, the full nonlinear equations of perfect-fluid motion show that the general character of these waves is preserved. The only difference is (from the normal component of the momentum equation, Equation 3.3a) that the magnitude of the total magnetic-field vector must remain constant, with the result that an oscillation can be created only if the disturbance vectors rotate about the undisturbed direction of the magnetic field. This has the further consequence that finite amplitude Alfvén waves propagate only along the magnetic-field lines, as expected from the conception of them as lines of tension. Sawyer, Scott, and Stratton[28] have observed finite amplitude Alfvén waves experimentally, but they make a statement indicating unawareness that these rotating waves have a simple theoretical explanation.

c. Magnetoacoustic waves: Under the assumption that $s' = 0$, consider the four scalar equations, Equation 3.32, the inner product of Equation 3.33 with $k_i$ and then with $B_i$, and the inner product of Equation 3.34 with $B_i$. These equations may be written

$$\begin{bmatrix} -\omega_0 & \rho & 0 & 0 \\ k^2 a^2 & -\rho\omega_0 & k^2/\mu & 0 \\ a^2(kB) & 0 & 0 & -\rho\omega_0 \\ 0 & B^2 & -\omega_0 & -(kB) \end{bmatrix} \begin{vmatrix} \tilde{\rho}' \\ (k\tilde{v}') \\ (B\tilde{B}') \\ (B\tilde{v}') \end{vmatrix} = 0 \qquad (3.42)$$

A nontrivial solution is obtained if the determinant of Equations 3.42 vanishes. This condition gives the following well-known equation for the eigenvalues of magnetoacoustic waves:

$$\omega_0^4 - k^2(a^2 + b^2)\omega_0^2 + k^2 a^2 (kb)^2 = 0 \qquad (3.43)$$

in which $b = B/\sqrt{\mu\rho}$. Equation 3.43 was also obtained in Chapter 2 (Equation 2.19) from the condition for a point of maximum entropy on the Rayleigh line.

From the first two of Equations 3.42,

$$(b\tilde{b}') = \left(\frac{\omega_0^2}{k^2} - a^2\right) \frac{\tilde{\rho}'}{\rho} \tag{3.44}$$

After substitution of Equation 3.44 and $(k\tilde{v}')$ from 3.32 into 3.33 and 3.34, the eigenvector of the magnetoacoustic wave becomes

$$\tilde{v}_i' = \frac{\omega_0}{k^2}\left[\frac{\omega_0^2 k_i - k^2(bk)b_i}{\omega_0^2 - (bk)^2}\right]\frac{\tilde{\rho}'}{\rho} \tag{3.45}$$

$$\tilde{b}_i' = \frac{\omega_0^2}{k^2}\left[\frac{k^2 b_i - (bk)k_i}{\omega_0^2 - (bk)^2}\right]\frac{\tilde{\rho}'}{\rho} \tag{3.46}$$

$$\tilde{S}' = 0$$

Thus if $\rho' \neq 0$ and Equation 3.40 does not hold, one obtains a family of waves which have both longitudinal and transverse components, showing the anisotropic character of the medium.

Phase Velocity. The bracketed expression 3.31 determines the amplitude at a given point on a wave. If it is constant to a particular observer, that observer must be traveling at a velocity which has a component in the direction of $\vec{k}$ of magnitude

$$v_p = \frac{\omega}{k} \tag{3.47}$$

called the phase velocity of the wave. The quantity $v_p$ is not a vector, but it can be made into a vector by arbitrarily defining

$$\vec{v}_p = \frac{\omega}{k}\frac{\vec{k}}{k} \tag{3.48}$$

This will be a useful concept if it is not misunderstood.

We can obtain the transformation properties of $\vec{v}_p$ by noting that the quantity $(k_i x_i - \omega t)$ must be the same to all observers; i.e., it is an invariant. Hence,

$$k_i x_i - \omega t = k_i' x_i' - \omega' t'$$

in which the primed reference frame is connected with the original one by the Galilean transformation,

$$x_i' = x_i - v_i t, \quad t' = t$$

Substituting back, we can see that

$$\left.\begin{aligned} k_i &= k_i' \\ \omega &= \omega' + k_i v_i \end{aligned}\right\} \qquad (3.49)$$

Thus the vector $\vec{k}$, like the length of a rod, is invariant under a Galilean transformation, and also a formula is obtained for the Doppler frequency shift.

From Equations 3.48 and 3.49 it is clear that the phase velocity in the primed coordinate system is

$$\vec{v}_p' = \frac{\omega'}{k^2}\,\vec{k}$$

Thus, $\vec{v}_p$ does not transform like an ordinary velocity at all; rather, its direction remains fixed, while its magnitude changes in proportion to the frequency shift.

The magnitudes of the phase velocities of small-amplitude waves in a magnetohydrodynamic fluid can now easily be found from the following formulas:

For entropy waves:
(Equation 3.37)
$$v_p = v_i \frac{k_i}{k} \qquad (3.50)$$

For Alfvén waves:
(Equations 3.40 and 3.37)
$$v_p = (v_i \pm b_i)\frac{k_i}{k} \qquad (3.51)$$

For magnetoacoustic waves:
(Equations 3.43 and 3.37)

$$\left(v_p - v_i\frac{k_i}{k}\right)^4 - (a^2 + b^2)\left(v_p - \frac{v_i k_i}{k}\right)^2 + a^2\left(\frac{k_i}{k}\,b_i\right)^2 = 0 \qquad (3.52)$$

Group Velocity. The group velocity is defined by the formula

$$v_{g_i} = \frac{\partial\omega}{\partial k_i} \qquad (3.53)$$

Formula 3.53 is usually derived as the velocity of a wave packet, which is composed of waves of all frequencies; however, here the concept is used in connection with plane monochromatic waves in an anisotropic medium. In the latter case, Lighthill[29] has shown by a very general argument that $\vec{v}_g$ is the velocity of energy propagation, which is the reason for its importance. (In the former case, this interpretation is well accepted.) We shall demonstrate this fact below specifically for the waves of interest here. Then, in the next few paragraphs we give an interpretation of the group velocity of magnetoacoustic waves in terms of the point-source or Huygens diagram.

By using Formula 3.53, the group velocities of the waves considered here can easily be obtained from Formulas 3.37, 3.40, and 3.43. They are as follows:

Entropy wave: $$v_{g_i} = v_i \qquad (3.54)$$

Alfvén wave: $$v_{g_i} = v_i \pm b_i \qquad (3.55)$$

Magnetoacoustic wave:

$$v_{g_i} = v_i + \frac{\omega_0^4 k_i - k^4 a^2 (kb)\, b_i}{\omega_0 k^2 \left[2\omega_0^2 - (a^2 + b^2)k^2\right]} \qquad (3.56)$$

As pointed out by Lighthill,[29] the component of the group velocity in the direction of the phase velocity is the phase velocity. This result is immediately obvious for entropy and Alfvén waves, and also may be seen to hold for magnetoacoustic waves by taking into account the identity

$$\omega_0^2 \left[2\omega_0^2 - (a^2 + b^2)k^2\right] = \omega_0^4 - k^2 a^2 (kb)^2 \qquad (3.57)$$

which follows directly from Equation 3.43.

The Velocity of Energy Propagation. The energy density and energy flux in perfect-fluid magnetohydrodynamics can be lifted directly from Equation 3.4a. They are as follows:

$$\text{Energy density} \equiv \rho U^* = \rho\left(U + \frac{v^2}{2}\right) + \frac{B^2}{2\mu} \qquad (3.58)$$

$$\text{Energy flux} \equiv F_i^* = \rho v_i\left(h + \frac{v^2}{2}\right) + \frac{B^2}{\mu} v_i - \frac{B_k v_k}{\mu} B_i \qquad (3.59)$$

The velocity of energy propagation of small-amplitude waves is then simply

$$v_{E_i} = \frac{F_i^{*\prime}}{(\rho U^*)'} \tag{3.60}$$

in which $F_i^{*\prime}$ and $(\rho U^*)'$ are the portions of $F_i^*$ and $\rho U^*$ due to small-amplitude waves. In particular, $F^{*\prime}$ is the energy flux of the wave with respect to the fluid. The quantities $F_i^{*\prime}$ and $(\rho U^*)'$ are found below by a procedure given by Landau and Lifshitz.[18]

First, consider the energy density of small-amplitude waves. The internal energy per unit volume can be expanded as follows:

$$\rho U = (\rho U)_0 + \left.\frac{\partial(\rho U)}{\partial \rho}\right]_0 \rho' + \frac{1}{2}\left.\frac{\partial^2(\rho U)}{\partial \rho^2}\right]_0 \rho'^2 + \cdots$$

In the present case of isentropic flow, application of well-known thermodynamic relations reduces this expression to

$$\rho U = (\rho U)_0 + h_0\rho' + \frac{1}{2}\left(\frac{a^2}{\rho}\right)_0 \rho'^2$$

The first term is the internal-energy density of the undisturbed fluid, and hence is dropped since only the energy density of the wave is desired. The fact that the second term also contributes nothing to the energy density of the wave can be seen by noting that the time average of the fluctuations of mass density due to the wave must be zero, because fluid is neither added nor removed by the wave. The third term, the square of an oscillatory quantity, has a nonzero mean and therefore must contribute to the energy density of the wave. The expansion of the magnetic part of Equation 3.58 is

$$\frac{B_0^2}{2\mu} + \frac{B_{0i}B_i'}{\mu} + \frac{B'^2}{2\mu}$$

The first and second terms can be discarded by using arguments similar to those used above. In the disposition of the second term, the argument depends on the well-known fact[30] that in an infinitely conducting fluid, the magnetic-field lines move with the fluid. This being the case, a change in magnetic flux in a given volume corresponds to a change in fluid mass in that volume. Since the time average of the latter change is zero, there can be no contribution to the energy density from the first-order magnetic term. As the kinetic-energy part of Equation 3.58 is already a square of a perturbation ($v = v'$), it is included in toto. The consequence of this discussion is that the energy density of a small-amplitude wave takes the form

$$(\rho U^*)' = \frac{1}{2}\frac{a^2}{\rho}\rho'^2 + \frac{1}{2}\rho b'^2 + \frac{1}{2}\rho v'^2 \tag{3.61}$$

in which the subscripts zero have been dropped and $b'^2 = B'^2/\mu\rho$.

Second, consider the energy flux of small-amplitude waves. The expansion of the specific enthalpy in isentropic flow can be expressed as

$$h = h_0 + \left(\frac{\partial h}{\partial \rho}\right)_s \rho' + \cdots = h_0 + \frac{a^2}{\rho}\rho' + \cdots$$

Consider the first term, $\rho v_i' h_0$, which will appear in Equation 3.59 after the above form of h has been substituted. The integral of $\rho v_i'$ over any surface in the fluid is the total mass flux through that surface. Since conservation of mass requires the time average of this quantity to be zero, the term $\rho v_i' h_0$ contributes nothing to the energy flux of the wave. Using this information, Equation 3.59 is expanded up to the second order and tentatively written

$$F_i^{*'} = \rho_0\left[\frac{a_0^2}{\rho_0} v_i'\rho' + 2(b_0 b')v_i' - (b_0 v')b_i' - (b'v')b_{i_0} + b_0^2 v_i' - (b_0 v')b_{i_0}\right]$$

The first four terms in the above expression are of second order in the perturbations, and hence have nonzero time averages. The last two terms are of first order; however, as will now be shown, they also have nonzero averages. They may be written

$$\rho_0 b_0^2 v_i' - \rho_0 (b_0 v')b_{i_0} = b_0^2 \rho v_i' - (b_0 \rho v')b_{i_0} - b_0^2 \rho' v_i' + \rho'(b_0 v')b_{i_0}$$

As shown for the case of the mechanical part of $F^*$, the integral of $\rho v_i'$ over a surface in the fluid has a zero time average. Thus, the first two terms on the right side of the above expression must be dropped, with the result that the expression for $F^{*'}$ becomes

$$F_i^{*'} = \rho \left[ \frac{a^2}{\rho} \rho' v_i' + 2(bb')v_i' - (bv')b_i' - (b'v')b_i - \frac{b^2}{\rho} \rho' v_i' + \frac{\rho'}{\rho} (bv')b_i \right] \tag{3.62}$$

in which the subscripts zero have been dropped because there is no further cause for ambiguity.

a. The entropy wave: For this case, Equations 3.38 show that $v_i' = b_i' = 0$. Hence, $F_i^* = 0$, but from Equation 3.61, $(\rho U^*)' \neq 0$. Since from Equation 3.60, $v_E = 0$, the entropy wave is a disturbance which moves with the fluid, and its total velocity of energy propagation is just the fluid velocity. This is also the group velocity of the entropy wave given by Equation 3.54.

b. The Alfvén wave: For this case, $\rho' = 0$ and $\tilde{v}_i' = \mp \tilde{b}_i'$ (Equation 3.41). Substituting these values into Equations 3.61 and 3.62 and taking into account that the mean value of the square of a harmonic function is one half the maximum, one obtains the time averages of energy density and energy flux:

$$\left.\begin{aligned} \langle U^{*'} \rangle &= \tfrac{1}{2}\rho \tilde{b}'^2 \\ \langle F_i^{*'} \rangle &= \mp \frac{\rho}{2} \left[ (b\tilde{b}')b_i' - \tilde{b}'^2 b_i \right] = \pm \tfrac{1}{2}\rho \tilde{b}'^2 b_i \end{aligned}\right\} \tag{3.63}$$

if the property of the Alfvén wave, $(b\tilde{b}') = 0$, is taken into account. Hence,

$$v_{Ei} = \frac{\langle F_i^{*'} \rangle}{\langle U^{*'} \rangle} = \pm b_i \tag{3.64}$$

which, if $v_i$ is added, agrees with Equation 3.55. Thus, in a reference frame at rest in the fluid, the energy of Alfvén waves propagates along the field lines. Furthermore, the equipartition between the magnetic- and kinetic-energy densities of the wave, found in deriving Equation 3.63, should be noted.

c. The magnetoacoustic wave: Much more algebra is required to find the formula for the energy-propagation velocity in this case. First, consider the kinetic-energy density given by Equation 3.61. By taking the inner product of Equation 3.45 with itself,

$$\tfrac{1}{2}\rho \tilde{v}'^2 = \frac{\omega_0^2 \tilde{\rho}'^2}{2k^4 \rho} \left[ \frac{\omega_0^2 k^2 (\omega_0^2 - (bk)^2) - k^2 (bk)^2 (\omega_0^2 - b^2 k^2)}{(\omega_0^2 - (bk)^2)^2} \right].$$

But Equation 3.43 can be written

$$\omega_0^2(\omega_0^2 - k^2b^2) = k^2a^2(\omega_0^2 - (kb)^2) \tag{3.65}$$

This allows one of the factors in $\frac{1}{2}\rho\tilde{v}'^2$ to be canceled, with the result,

$$\frac{1}{2}\rho\tilde{v}'^2 = \frac{1}{2\rho k^2}\left[\frac{\omega_0^4 - k^2a^2(bk)^2}{\omega_0^2 - (bk)^2}\right]\tilde{\rho}'^2 \tag{3.66}$$

With the help of the following identity, derived from Equation 3.65,

$$\omega_0^2[k^2b^2 - (kb)^2] = (\omega_0^2 - (bk)^2)(\omega_0^2 - k^2a^2) \tag{3.67}$$

the inner product of Equation 3.46 with itself can be put in the form

$$\tilde{b}'^2 = \frac{\omega_0^2}{k^2}\left[\frac{\omega_0^2 - k^2a^2}{\omega_0^2 - (bk)^2}\right]\frac{\tilde{\rho}'^2}{\rho^2} \tag{3.68}$$

Then,

$$\frac{1}{2}\rho\left(\frac{a^2}{\rho^2}\tilde{\rho}'^2 + \tilde{b}'^2\right) = \frac{1}{2\rho k^2}\left[\frac{\omega_0^4 - k^2a^2(bk)^2}{\omega_0^2 - (bk)^2}\right]\tilde{\rho}'^2 \tag{3.69}$$

which agrees exactly with Equation 3.66, thus showing equipartition between kinetic energy and the sum of magnetic and pressure energy, i.e., potential energy. Substituting Equations 3.66 and 3.69 into 3.61, and using Equation 3.57 to modify the form of the result, we can write the time average of the energy density of magnetoacoustic waves as follows:

$$\langle U^{*\prime}\rangle = \frac{\omega_0^2}{2k^2\rho}\left[\frac{2\omega_0^2 - k^2(a^2 + b^2)}{\omega_0^2 - (bk)^2}\right]\tilde{\rho}'^2 \tag{3.70}$$

Second, consider the energy flux given by Equation 3.62. With the help of Identities 3.65 and 3.67,

$$(b\tilde{b}')\tilde{v}_i' = \frac{\omega_0}{k^4}\left[\frac{(\omega_0^2 - k^2a^2)[\omega_0^2k_i - k^2(bk)b_i]}{\omega_0^2 \quad (bk)^2}\right]\frac{\tilde{\rho}'^2}{\rho^2} \tag{3.71}$$

$$(\tilde{bv}')\tilde{b}_i' = \omega_0 \frac{(bk)}{k^2} a^2 \left[\frac{k^2 b_i - (bk)k_i}{\omega_0^2 - (bk)^2}\right] \frac{\tilde{\rho}'^2}{\rho^2} \tag{3.72}$$

$$(\tilde{b}'\tilde{v}')b_i = -\frac{\omega_0 (bk)}{k^2} \left[\frac{\omega_0^2 - k^2 a^2}{\omega_0^2 - (bk)^2}\right] \frac{\tilde{\rho}'^2}{\rho^2} b_i \tag{3.73}$$

Then, after substitution of the above three expressions and Equations 3.45 and 3.46 into Equation 3.62, the time average of the energy flux for magnetoacoustic waves reduces to

$$<F_i^{*\prime}> = \frac{\omega_0}{2k^4} \left[\frac{\omega_0^4 k_i - k^4 a^2 (bk) b_i}{\omega_0^2 - (bk)^2}\right] \frac{\tilde{\rho}'^2}{\rho} \tag{3.74}$$

Finally, dividing Equation 3.74 by Equation 3.70, we see that the energy of magnetoacoustic waves propagates at the following velocity with respect to the fluid:

$$v_{E_i} = \frac{<F_i^{*\prime}>}{<U^{*\prime}>} = \frac{1}{\omega_0 k^2} \left[\frac{\omega_0^4 k_i - k^4 a^2 (bk) b_i}{2\omega_0^2 - (a^2 + b^2) k^2}\right] \tag{3.75}$$

Comparison of Equations 3.75 and 3.56 confirms the interpretation of group velocity as the energy-propagation velocity of the wave.

The Point-Source Diagram. To strengthen the concept of group velocity, we now consider its relation to the diagram of the shape of a disturbance propagating from a point disturbance. This diagram is referred to variously as the Huygens diagram or the Friedrichs diagram. For magnetohydrodynamics, it has been discussed by Friedrichs,[3] Grad,[31] Sears,[32] and others; however, in all papers known to this author it is constructed graphically from the phase-velocity diagram. In this section we derive a simple analytical formula for the point-source diagram, and show it to be exactly the polar diagram of the group velocity.

Consider Formula 3.52 for the phase velocity of magnetoacoustic waves. Take a coordinate system in which the undisturbed fluid is at rest ($v_i = 0$), and let

$$r = \frac{v_\rho}{\sqrt{ab}}, \qquad \chi = \frac{a^2 + b^2}{2ab} \tag{3.76}$$

Then Equation 3.52 can be written

$$r^4 - 2\chi r^2 + \cos^2\theta = 0 \tag{3.77}$$

in which $\theta$ is the angle between the magnetic field and wave-propagation vectors. Solving for $r$, we obtain the well-known equation for the polar phase-velocity diagram of fast and slow magnetoacoustic waves:

$$r = \sqrt{\chi \pm \sqrt{\chi^2 - \cos^2\theta}} \qquad (3.78)$$

It is plotted in Figure 3.3 for a particular value of $\chi$.

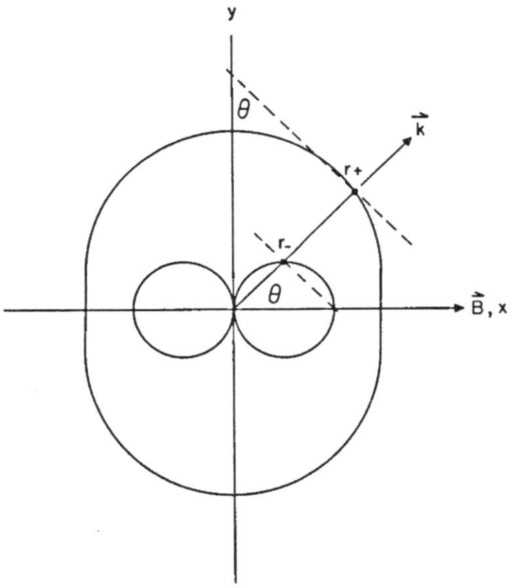


Figure 3.3. Phase-velocity diagram for magnetoacoustic waves

Consider a self-similar pulse propagated from a point source at the origin. The wave front generated by this pulse moves with different velocities in different directions, as shown in the figure. If the wave front (of infinitesimal amplitude) is considered to be a superposition of plane waves, the positions of the fronts of fast and slow waves propagated in the direction $\theta$ will be, after unit time, the dotted lines perpendicular to $\vec{k}$ in Figure 3.3. Hence, the entire wave front of the self-similar pulse will be the envelope of the family of straight lines like the dotted lines in the figure obtained by varying $\theta$. The equation of these lines is easily seen to be

$$y = -\cot\theta\, x + \frac{r}{\sin\theta}$$

or

$$f(x, y, \theta) = y \sin\theta + x\cos\theta - r(\theta) = 0 \qquad (3.79)$$

For each line $f = 0$, defined by a particular $\theta$, there is a point of the envelope which satisfies $f = 0$ for that $\theta$. Thus, the equation of the envelope may be written parametrically as

$$y = y(\theta), \quad x = x(\theta)$$

and for it,

$$f[x(\theta),\ y(\theta),\ \theta] = 0$$

holds. The derivative of f with respect to θ is

$$\frac{\partial f}{\partial x}\frac{dx}{d\theta} + \frac{\partial f}{\partial y}\frac{dy}{d\theta} + \frac{\partial f}{\partial \theta} = 0 \qquad (3.80)$$

which, solved for dy/dx, must give the slope of the envelope. But, for a given θ, the slope of the line $f = 0$ and the slope of the envelope are by definition equal. The former is obtained by differentiating Equation 3.79 with θ fixed; thus,

$$\frac{\partial f}{\partial x}\, dx + \frac{\partial f}{\partial y}\, dy = 0 \qquad (3.81)$$

After comparing Equations 3.80 and 3.81, it is clear that the slope of the line $f = 0$ and the slope of the envelope can agree only if

$$\frac{\partial f}{\partial \theta} = 0 = y \cos \theta - x \sin \theta - \frac{dr}{d\theta} \qquad (3.82)$$

Thus, the envelope is obtained by eliminating θ between the equations $f = 0$ and $\partial f/\partial\theta = 0$. (This method of finding the envelope is well known, and is called by Ince[33] a "singular solution.")

Because of the transcendental nature of Equations 3.79 and 3.82, θ cannot be eliminated between them; however, convenient formulas for the polar coordinates of the point-source diagram (R, ϕ) can be obtained in terms of the polar coordinates of the phase-velocity diagram (r, θ). In Equations 3.79 and 3.82, x and y are the Cartesian coordinates of the envelope. Let

$$y = R \sin \phi$$

$$x = R \cos \phi$$

Then, Equations 3.79 and 3.82 become

$$\left.\begin{aligned} r(\theta) &= R \cos(\phi - \theta) \\ r'(\theta) &= \frac{\sin 2\theta}{4r(r^2 - \chi)} = R \sin(\phi - \theta) \end{aligned}\right\} \qquad (3.83)$$

in which $r'(\theta)$ has been found from Equation 3.77. Squaring and adding Equations 3.83 and substituting for $(r^2 - \chi)^2$ from Equation 3.78, we get

$$R^2 = r^2 + \frac{\sin^2 2\theta}{16r^2(\chi^2 - \cos^2\theta)} \tag{3.84}$$

If we divide one of Equations 3.83 by the other and take Equation 3.77 into account, we obtain the following equation:

$$\tan\phi = \frac{r^4}{r^4 - 1}\tan\theta \tag{3.85}$$

Thus, for a given point on the phase-velocity diagram, Equations 3.84 and 3.85 give the corresponding point on the point-source diagram. For the case $\chi = 0.75\sqrt{2}$, the first quadrant of each of the two diagrams is plotted in Figure 3.4, in which the correspondence between points is indicated.

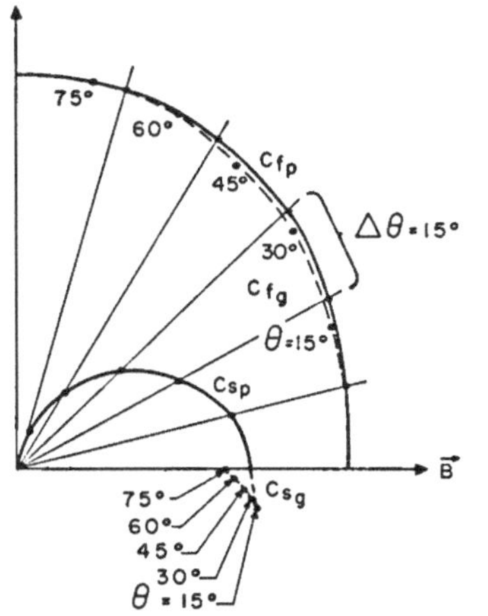


Figure 3.4. Phase and group velocity diagram

Now consider the similar polar-coordinate plot of the group velocity in a reference frame at rest with respect to the fluid. Using the definition of phase velocity given by Equation 3.47, we can write Equation 3.56, in terms of the notation in 3.76, in vector notation as

$$\frac{\vec{v}_g}{\sqrt{ab}} = \vec{R}_g$$

$$= \frac{r^4 \dfrac{\vec{k}}{k} - \cos\theta \dfrac{\vec{b}}{b}}{2r(r^2 - \chi)}$$

Using the definition of $\phi_g$ shown in Figure 3.5, we find that the Cartesian components of $\vec{R}_g$ are

$$R_g \cos\phi_g = \frac{(r^4 - 1)\cos\theta}{2r(r^2 - \chi)}$$

$$R_g \sin\phi_g = \frac{r^4 \sin\theta}{2r(r^2 - \chi)}$$

Dividing the second by the first, we have

$$\tan \phi_g = \frac{r^4}{r^4 - 1} \tan \theta \tag{3.86}$$

Squaring and adding, we can write the following result:

$$R_g^2 = \frac{(r^4 + \cos^2 \theta)^2 - 4r^4 \cos^2 \theta + \cos^2 \theta(1 - \cos^2 \theta)}{4r^2(r^2 - \chi)^2}$$

which, with the help of Equation 3.77 and 3.78, becomes

$$R_g^2 = r^2 + \frac{\sin^2 2\theta}{16r^2(\chi^2 - \cos^2 \theta)} \tag{3.87}$$

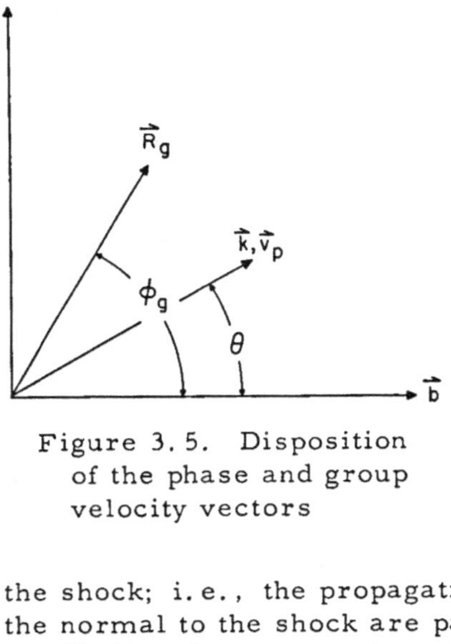


Figure 3.5. Disposition of the phase and group velocity vectors

The agreement between Equations 3.84 and 3.87 and between Equations 3.85 and 3.86 indicates that the wave front from a point source in any particular direction propagates with the group velocity in that direction. These developments now show clearly the meaning of the concept of group velocity in an anisotropic medium.

## 3. Stability with Respect to Normal Small Disturbances

In this section, the shock is disturbed by a plane monochromatic wave of any of the possible types normally incident upon the shock; i.e., the propagation vector of the incident wave and the normal to the shock are parallel. First, we present the small-disturbance equations and boundary conditions for this case and show that they split in two groups, one for magnetoacoustic- and entropy-wave disturbances and the other for Alfvén-wave disturbances. Then, we set up the problem of finding the amplitudes of divergent waves in each group and explore its consequences in terms of shock stability. The result will be that the flow domain is separated into "evolutionary" and "nonevolutionary" regions. Finally, we discuss the behavior of the amplitudes of reflected and refracted Alfvén waves near the boundary of these regions. Except for the treatment of diverging wave amplitudes, the work of this section is due to Akhiezer, Lyubarskii, and Polovin[34] and to Syrovatskii.[35]

Small-Disturbance Equations and Boundary Conditions. Equations 3.11 through 3.15 are valid for arbitrary small disturbances around a flow condition in which the steady-state variables are constant. If the x-axis is taken along the normal to the shock, the present case can be characterized by the fact that the disturbances are functions of x and t only. Then without loss of generality, the x-y plane on both sides of the shock can be taken parallel to the plane of the shock normal and the magnetic-field and velocity vectors. Thus $v_z = B_z = 0$. Equations 3.11 through 3.15 then split into the following two sets:

$$\left(\delta_{ij}\frac{\partial}{\partial t} + {}^5X_{ij}\frac{\partial}{\partial x}\right){}^5\xi_j' = 0 \qquad (i, j = 1, 2, \cdots, 5) \tag{3.88}$$

in which

$${}^5X_{ij} = \begin{bmatrix} v_x & \rho & 0 & 0 & 0 \\ a^2/\rho & v_x & 0 & b_y & p_s/\rho \\ 0 & 0 & v_x & -b_x & 0 \\ 0 & b_y & -b_x & v_x & 0 \\ 0 & 0 & 0 & 0 & v_x \end{bmatrix}; \quad \left|{}^5\xi_j'\right| = \begin{vmatrix} \rho' \\ v_x' \\ v_y' \\ b_y' \\ s' \end{vmatrix}$$

and

$$\left(\delta_{ij}\frac{\partial}{\partial t} + {}^2X_{ij}\frac{\partial}{\partial x}\right){}^2\xi_j' = 0 \qquad (i, j = 1, 2) \tag{3.89}$$

in which

$${}^2X_{ij} = \begin{bmatrix} v_x & -b_x \\ -b_x & v_x \end{bmatrix}; \quad \left|{}^2\xi_j'\right| = \begin{vmatrix} v_z' \\ b_z' \end{vmatrix}$$

Moreover, the $x^{th}$ of Equations 3.13 and the Equation 3.15 together show that $B_x' = \text{const}$. But $B_x'$ is in general an oscillatory quantity; therefore, $B_x' = 0$.

For the present case, we have already indicated that the boundary conditions (Equations 3.22 through 3.29) likewise split into two groups. Under the conditions imposed above, the first group becomes

$$\left\{ M\,\xi \right\} = 0 \tag{3.90}$$

in which M is the $5 \times 6$ matrix:

$$\begin{bmatrix} -\rho & v_x & \rho & 0 & 0 & 0 \\ 0 & v_x^2 + a^2 & 2\rho v_x & 0 & \rho b_y & p_s \\ -\rho v_y & v_x v_y & \rho v_y & \rho v_x & -\rho b_x & 0 \\ -\rho \bar{U}^* & v_x\left(h + \frac{v^2}{2} + a^2\right) & \rho \bar{U}^* + \rho v_x^2 + p^* & \rho(v_x v_y - b_x b_y) & \rho(2v_x b_y - v_y b_x) & (\rho T + p_s)v_x \\ -\rho^{1/2} b_y & 0 & \rho^{1/2} b_y & -\rho^{1/2} b_x & \rho^{1/2} v_x & 0 \end{bmatrix}$$

and $\xi$ is the $6 \times 1$ column matrix:

$$\begin{vmatrix} \zeta_t \\ {}^5\xi_j' \end{vmatrix}$$

consisting of the perturbed shock velocity $\zeta_t$ and the perturbed quantities in Equation 3.88. The braces, $\{\ \}$, indicate the jump in the enclosed quantity in crossing the shock.

We obtain the second group of boundary conditions from Equations 3.25 and 3.28 by setting $v_z = B_z = 0$. They are

$$\left\{ \begin{bmatrix} \rho v_x & -\rho b_x \\ -\rho^{1/2} b_x & \rho^{1/2} v_x \end{bmatrix} \begin{vmatrix} v_z' \\ b_z' \end{vmatrix} \right\} = 0 \tag{3.91}$$

Magnetoacoustic Incident Wave. In this paragraph we assume the wave incident on the shock to be a magnetoacoustic wave; i.e., oscillations in the velocity and magnetic-field vectors are in the plane formed by these vectors and the normal to the shock. In this case the disturbance is governed by Equations 3.88 and boundary conditions of 3.90. The latter show that since $v_z'$ and $b_z'$ do not appear, all divergent waves must be either magnetoacoustic or entropy waves, and no Alfvén waves can arise.

Under the assumption that the waves are plane and monochromatic, Equations 3.88 have already been solved on pages 42 - 45. The results were expressed in the form of eigenvalues and their corresponding eigenvectors. For the set of equations corresponding to Equations 3.88, there were five solutions — each corresponding to a different wave — and their sum is the general solution given in terms of five arbitrary constants. One solution was the entropy wave, which has an eigenvalue $\omega_0 = 0$ and an eigenvector given by Equation 3.38. The other four solutions are the fast and slow sound (magnetoacoustic) waves moving in two opposite directions. Their eigenvalues are found from Equation 3.43, and the corresponding eigenvectors from 3.45 and 3.46. For the work of this section the most important results of Section 2 are the eigenvalues of the waves. They were found to be expressible in terms of the phase velocities, which, for entropy and magnetoacoustic waves, are given respectively by Equations 3.50 and 3.52. Since $\vec{k}$ is directed along the x-axis in the present case, the phase velocity of the entropy wave is $v_p = v_x$. There are four magnetoacoustic waves, which, in terms of notation defined in Equation 2.19, have the following phase velocities with respect to the shock:

$$v_p = v_x \pm c_f,\ v_x \pm c_s \tag{3.92}$$

It has just been shown that in the flow field on each side of the shock, five types of waves (each, at this point, having an arbitrary amplitude) may be possible. Some of them, however, will be eliminated because of their relation to the normal component of the flow velocity $v_x$. For example, since the entropy wave moves with the fluid, it can never exist on the upstream side of the shock, but always exists on the downstream side. The most convenient way to show the types of waves possible is by a diagram of the type given in Figure 3.6.

To understand Figure 3.6, assume that at $t = 0$ a shock exists with an arbitrary set of values $v_{x_1}$ and $v_{x_2}$. Then an incident fast or slow sound wave comes from, say, upstream and hits the shock. In each of the nine regions shown on Figure 3.6, the types of waves that will be reflected or refracted from the shock are indicated by arrows of appropriate relative length and by descriptive symbols. Thus, on the downstream side (to the right) the waves $v_x + c_f$ and $v_x + c_s$ and the entropy wave always appear, while on the upstream side these three waves never appear. The other waves indicated by Equation 3.92 appear only as shown. The numeral in the lower left-hand corner of each region, shown for convenience, is the total number of diverging waves in that region.

For each wave shown in Figure 3.6, there is an eigenvector which gives a relationship between the variables $\tilde{\rho}'$, $\tilde{s}'$, $\tilde{v}_x'$,

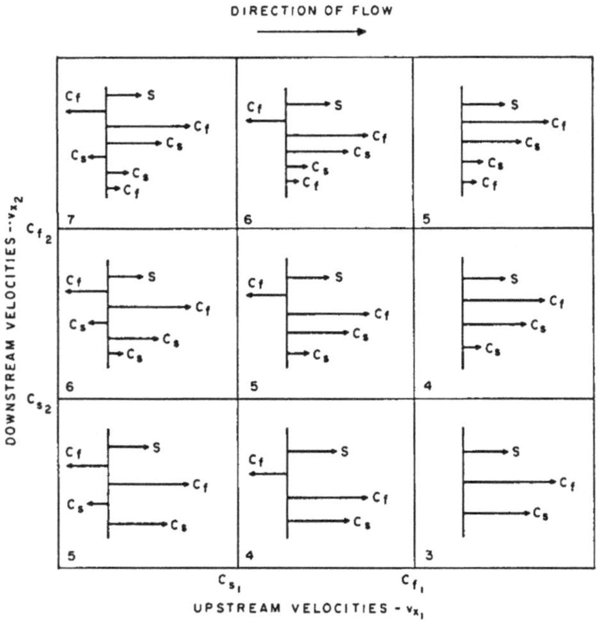


Figure 3.6. Possible divergent waves resulting from a magnetoacoustic incident wave

$\tilde{v}_y'$, $\tilde{b}_y'$; however, the magnitude of the eigenvector, which determines the strength of the wave, is arbitrary. For a given incident wave, the strengths of the diverging waves must be found from the boundary conditions, Equations 3.90. After eliminating the perturbed shock velocity $\zeta_t$ from them, there are exactly four equations from which to determine the strengths of diverging waves. Hence, only in the two regions of Figure 3.6 in which there are just four diverging waves can the strengths be uniquely determined. When there are more than four diverging waves, an infinite number of solutions for their strengths can be found; in fact, nonzero solutions can be found even when the incident wave strength is zero. In this case, therefore, the shock can break up spontaneously by emitting magnetoacoustic and entropy waves. In comparing Figures 2.10 and 2.11 with Figure 3.6, it is interesting to note that the regions where spontaneous emission is possible are those corresponding to expansion shocks ($\beta < 1$), which do not satisfy the second law of thermodynamics.

In the one region of Figure 3.6 in which there are only three diverging waves, no solution at all can be found which will satisfy all of the boundary conditions. In this case spontaneous emission is not possible since, when the incident wave strength is zero, the boundary conditions are satisfied only if the divergent wave strengths are zero. Besides, this case corresponds to the region of Figure 2.11 in which the intermediate shocks lie. These shocks do satisfy the second law of thermodynamics. The fact that there is no solution to the small-disturbance problem is, however, significant. A specific steady-flow situation has been postulated to exist at $t = 0$. Then, the problem of propagation of small disturbances in this flow has been set up and has been found to have no solution. If such a flow did exist and were stable to an infinite train of normal Alfvén waves, it would certainly have a solution. Thus, we must conclude, tentatively at least, that the initial postulated flow cannot exist in the presence of small disturbances; if it did for an instant, it must break up immediately. This conclusion is strengthened by the discussion in Section 5.

As indicated in the introduction to this chapter, shocks for which a unique solution to the small-disturbance problem cannot be found have been called "nonevolutionary" in the Russian literature. The growth of disturbances in the neighborhood of nonevolutionary shocks cannot be traced, and if such discontinuities are formed by some means, the theory implies that they must break up immediately. Shocks which satisfy the inequalities

$$\left.\begin{aligned} c_{f_1} &< v_{x_1} < \infty \\ c_{s_2} &< v_{x_2} < c_{f_2} \end{aligned}\right\} \qquad (3.93)$$

or

$$\left.\begin{aligned} c_{s_1} &< v_{x_1} < c_{f_1} \\ 0 &< v_{x_2} < c_{s_2} \end{aligned}\right\} \qquad (3.94)$$

would be called evolutionary with respect to magnetoacoustic incident disturbances; however, a final classification of evolutionary and nonevolutionary shocks must be deferred until after the case of incident Alfvén waves has been discussed.

In this section, the shock has been perturbed by a plane monochromatic wave; but since an arbitrary infinitesimal disturbance is made up of an integral over plane waves, the results of this study must be the same as if an arbitrary initial disturbance had been assumed. The latter approach was taken in Reference 34, where the above results were found with somewhat greater elegance but at the sacrifice of some clarity.

Alfvén Incident Wave. If the disturbance incident on the shock has velocity and magnetic-field components perpendicular to the x-y plane only, the disturbance-wave components $v_z'$ and $b_z'$ are found from Equations 3.89. For this type of disturbance, Equations 3.91 give the boundary conditions. We have already solved Equations 3.89 for the case of plane monochromatic waves; the results were the eigenvalues given by Equations 3.40 and the nonzero components of the eigenvectors given by Equations 3.41.

The eigenvalues are expressed in terms of phase velocity by Equations 3.51, and in the present case they become

$$v_p = v_x \pm b_x \tag{3.95}$$

Just as in the preceding section, it is not always possible to have both of these waves on both sides of the shock. The ones which are possible are shown in Figure 3.7. In the present case, there are two boundary conditions (3.91); hence, the amplitudes of diverging waves can be found only in the regions of Figure 3.7 in which just two of them appear. That is, the shock is evolutionary only if

$$\left.\begin{array}{ll} b_{x_1} < v_{x_1} < \infty, & b_{x_2} < v_{x_2} < \infty \\ \text{or if} & \\ 0 < v_{x_1} < b_{x_1}, & 0 < v_{x_2} < b_{x_2} \end{array}\right\} \tag{3.96}$$

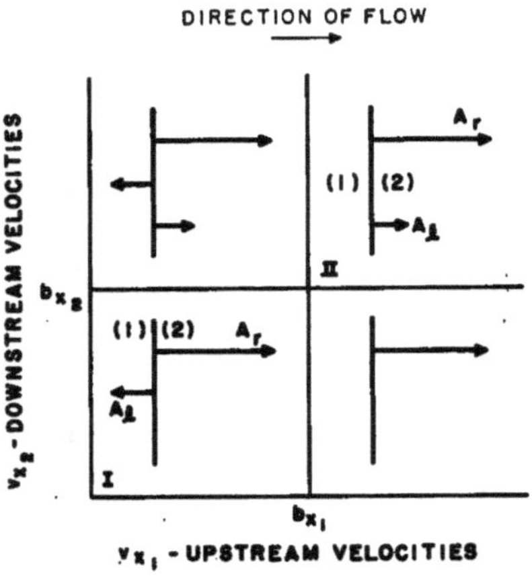


Figure 3.7. Possible divergent waves resulting from an Alfvén incident wave

Again, in the region in which there are more diverging waves than boundary conditions, spontaneous emission of Alfvén waves from the shock is possible, and this is part of the region of Figure 2.11 corresponding to expansion shocks. Note also that the region of Figure 3.7 which has only one diverging wave is the region of Figure 2.11 corresponding to intermediate shocks.

In an attempt to gain some insight into the above-described behavior, we shall now compute the amplitudes of the diverging Alfvén waves in the two evolutionary regions of Figure 3.7. Then we shall examine the behavior as the nonevolutionary region is approached. First, consider

Region I in Figure 3.7. Let an Alfvén wave for which $|b_x'| = 1$ strike the shock from the upstream side. Then, the total disturbance on the upstream side is the sum of the incident wave and the reflected wave. Thus,

$$b_{z_1}' = e^{i\omega\left(\frac{x}{v_{x_1} + b_{x_1}} - t\right)} + A_\ell e^{i\omega\left(\frac{x}{v_{x_1} - b_{x_1}} - t\right)}$$

and, from Equation 3.41,

$$v_{z_1} = - e^{i\omega\left(\frac{x}{v_{x_1} + b_{x_1}} - t\right)} + A_\ell e^{i\omega\left(\frac{x}{v_{x_1} - b_{x_1}} - t\right)} \qquad (3.97)$$

where $A_\ell$ is the amplitude of the reflected wave. Here it is assumed that $B_x$ is in the direction of the flow. On the downstream side the single refracted wave can be expressed as

$$\left.\begin{aligned} b_{z_2}' &= A_r e^{i\omega\left(\frac{x}{v_{x_2} + b_{x_2}} - t\right)} \\ v_{z_2}' &= -A_r e^{i\omega\left(\frac{x}{v_{x_2} + b_{x_2}} - t\right)} \end{aligned}\right\} \qquad (3.98)$$

The same value of $\omega$ is used throughout Equations 3.97 and 3.98 because of the obvious boundary condition that in a particular reference frame the frequencies of all diverging waves must be the same as the frequency of the incident wave. Now, Equations 3.97 and 3.98 with $x = 0$ are substituted into the boundary conditions of Equations 3.91. Then canceling the common factor $e^{-i\omega t}$, we have

$$\rho_1 v_{x_1}(-1 + A_\ell) - \rho_1 b_{x_1}(1 + A_\ell) = -\rho_2 v_{x_2} A_r \quad \rho_2 b_{x_2} A_r$$

$$\rho_1^{1/2} b_{x_1}(-1 + A_\ell) - \rho_1^{1/2} v_{x_1}(1 + A_\ell) = -\rho_2^{1/2} b_{x_2} A_r - \rho_2^{1/2} v_{x_2} A_r$$

from which

$$\left.\begin{aligned} A_\ell &= \left(\frac{1 - \sqrt{\frac{\rho_2}{\rho_1}}}{1 + \sqrt{\frac{\rho_2}{\rho_1}}}\right)\left(\frac{v_{x_1} + b_{x_1}}{v_{x_1} - b_{x_1}}\right) \\ A_r &= \left(\frac{2}{\frac{\rho_2}{\rho_1} + \sqrt{\frac{\rho_2}{\rho_1}}}\right)\left(\frac{v_{x_1} + b_{x_1}}{v_{x_2} + b_{x_2}}\right) \end{aligned}\right\} \tag{3.99}$$

In a similar manner the amplitudes of the two refracted waves in Region II of Figure 3.7 are found to be

$$\left.\begin{aligned} A_r &= \frac{1}{2}\left(\sqrt{\frac{\rho_1}{\rho_2}} + \frac{\rho_1}{\rho_2}\right)\left(\frac{v_{x_1} + b_{x_1}}{v_{x_2} + b_{x_2}}\right) \\ A_\ell &= \frac{1}{2}\left(\sqrt{\frac{\rho_1}{\rho_2}} - \frac{\rho_1}{\rho_2}\right)\left(\frac{v_{x_1} + b_{x_1}}{v_{x_2} - b_{x_2}}\right) \end{aligned}\right\} \tag{3.100}$$

Again, compare Figures 2.11 and 3.7. Equations 3.99 and 3.100 give the amplitudes of normal Alfvén waves reflected and refracted from slow and fast shocks, respectively. As $v_{x_1}$ approaches $b_{x_1}$ from Region I, the slow shock transforms into a "switch-off" shock, and as $v_{x_2}$ approaches $b_{x_2}$ from Region II, the fast shock transforms into a "switch-on" shock. It is evident that as these limits are approached, the divergent-wave amplitude $A_\ell$ approaches infinity; hence, the small-disturbance solution breaks down, and the shock cannot retain its original form.

Further insight into this phenomenon is achieved by considering an energy balance between the incident and divergent waves. One must take into account that the shock can put energy into the diverging waves in order to create an actual balance of energies. Then, the following conservation equation may be written:

$$F_I + F_s = F_\ell + F_r \tag{3.101}$$

in which $F_I$ is the energy flux of the input wave, $F_s$ is the energy per unit cross section per unit time added by the shock, and $F_\ell$ and $F_r$ are the magnitudes of the energy fluxes of the two divergent waves.

The energy density and the energy flux with respect to the fluid are given by Equations 3.63. We can write the energy flux in a coordinate system at rest with respect to the shock simply as the energy density multiplied by the velocity of the wave with respect to the shock; therefore, if we substitute the wave amplitudes and velocities from Equation 3.97 (for Region I) into Equation 3.101,

$$\tfrac{1}{2}\rho_1(v_{x_1} + b_{x_1}) + F_s = -\tfrac{1}{2}\rho_1 A_\ell^2(v_{x_1} - b_{x_1}) + \tfrac{1}{2}\rho_2 A_r^2(v_{x_2} + b_{x_2})$$

After substituting for $A_\ell$ and $A_r$ from Equation 3.99 and using the steady-state continuity equation, we can express the energy added by the shock in the form

$$F_s = \frac{(\rho_2 - \rho_1)v_{x_1}}{\left(1 + \sqrt{\dfrac{\rho_2}{\rho_1}}\right)^2}\left(\frac{v_{x_1} + b_{x_1}}{v_{x_2} + b_{x_2}}\right)\left(\frac{v_{x_2} - b_{x_2}}{v_{x_1} - b_{x_1}}\right) \qquad (3.102)$$

In order to get the correct dimensions for $F_s$, remember that the energy per unit mass of the incident wave was taken to be unity.

Equation 3.102 shows that in the region of Figure 2.11 corresponding to $3 \rightarrow 4$ (slow) shocks, $F_s > 0$. As the density ratio of slow shocks increases, $v_{x_1}$ approaches $b_{x_1}$, at which point the linear theory* predicts that an infinite amount of energy is put into the reflected wave by the shock. Now, from the physical point of view, consider the flow from a reference frame at rest with respect to the fluid ahead of the shock. Then, the shock moves toward the incident Alfvén wave with a velocity $v_{x_1}$. Let the shock velocity approach the Alfvén speed in the fluid ahead. Then the shock appears to act as a forcing function just below and approaching a natural frequency of the system. When the natural frequency is reached, it is natural to expect that the original configuration will be destroyed.

There is a more specific mechanical anology which, though not perfect, may help the reader to acquire a physical feeling for the above-described process. Imagine a long thin string

---

* Actually, the results here are more general than indicated. As discussed earlier, the Alfvén wave is a strictly linear phenomenon, and the only modification resulting if we drop the small-disturbance assumption is that the wave is three-dimensional and rotating. This lends further credence to the conclusion that the shock must break up as $v_{x_1}$ approaches $b_{x_1}$

being pulled through a smooth hole at a velocity $u$. Suppose that the hole is small enough so that it applies a friction force to the string and thereby changes the tension in it. Let the speed of propagation of small-amplitude waves, proportional to the square root of the tension, be $c_1$ on the tight side and $c_2$ on the loose side. Thus, $c_1 > c_2$. If one end of the string is oscillated laterally, small-amplitude waves move toward the hole and will be reflected and refracted there. In the two cases, $c_1 > c_2 > u$ and $u > c_1 > c_2$, two waves diverge from the hole; but in the case $c_1 > u > c_2$, only one wave diverges. There are two boundary conditions at the hole, one expressing the continuity of the deflection of the string, and one prescribing a change in slope due to the geometry of the hole. Thus, one can see an analogy to the shock problem. In the cases in which there are two diverging waves, one obtains a unique solution for the amplitudes because there are two waves and two boundary conditions; but in the intermediate case, one obtains no solution at all. The experiment in the intermediate case could be performed — conceptually at least. The fact that a linearized problem related to it has no solution can mean — it would seem — only that at some point small-amplitude theory breaks down. In this case, the configuration is said to be unstable; i.e., if one attempts to pull a long string through a hole, as described above, at a velocity intermediate between $c_1$ and $c_2$, one incites violent oscillations in the string.

An important and interesting fact to note is that the analysis indicates that the presence of a normal magnetic field alone can break up a pure gas-dynamical shock in a conducting fluid. This case is of particular interest since the effect of the magnetic field disappears completely in the steady-state shock equations. Thus the steady-state shock properties are exactly those of ordinary shocks, and in particular a shock layer is known to exist.[10] Only by studying the effect of Alfvén disturbances can the region of instability be found. This suggests that a process similar to that associated with the above-described analogy is a primary cause of breakup of intermediate magnetohydrodynamic shocks.

The region of instability (or, more precisely, nonevolutionarity) when $B_y = 0$ can easily be found by the following argument. Suppose that the point S in Figure 3.8 satisfies the Rankine-Hugoniot relations for ordinary gas-dynamic shocks. From the continuity equation $v_{x_2}/v_{x_1} = \rho_1/\rho_2$, each ray from the origin is a line of constant density ratio. But the ratio of Alfvén speeds is

$$\frac{b_{x_2}}{b_{x_1}} = \sqrt{\frac{\rho_1}{\rho_2}}$$

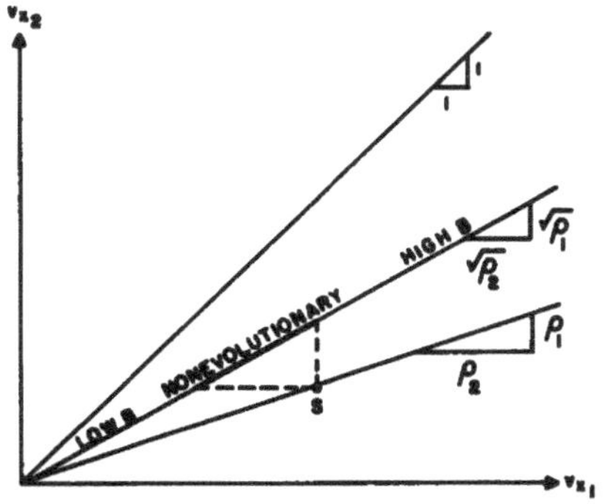


Figure 3.8. Region of nonevolutionarity of gas-dynamical shocks in a conducting fluid with a normal magnetic field

Thus, for the given shock S having a particular density ratio $\rho_2/\rho_1$, the locus of Alfvén speeds must be the ray from the origin of slope $\sqrt{\rho_1/\rho_2}$. Since $\rho_1/\rho_2 < 1$ for compressive shocks,

$$\frac{\rho_1}{\rho_2} < \sqrt{\frac{\rho_1}{\rho_2}}$$

so the locus of Alfvén speeds always lies above point S. After comparing Figure 3.8 with Figure 3.7, it is clear that point S lies in a nonevolutionary region whenever $B_x$ is such that the Alfvén speeds lie on the heavily drawn line segment of Figure 3.8. Thus, the shock must break up when the following pair of conditions hold:

$$\left.\begin{aligned} v_{x_1} &> b_{x_1} \\ v_{x_2} &< b_{x_2} \end{aligned}\right\} \qquad (3.103)$$

Summary of the Effect of Normal Disturbances. If Figures 3.6 and 3.7 are superimposed, and Inequalities 2.51 are taken into account, it is seen that oblique magnetohydrodynamic shocks are evolutionary only if

$$c_{f_1} < v_{x_1} < \infty, \qquad b_{x_2} < v_{x_2} < c_{f_2}$$

or if

$$c_{s_1} < v_{x_1} < b_{x_1}, \qquad 0 < v_{x_2} < c_{s_2}$$

These are the regions of Figure 2.11 in which $1 \rightarrow 2$ (fast) and $3 \rightarrow 4$ (slow) shocks lie. The conclusion is then that only the solid portions of the shock adiabatic of Figure 2.11 are evolutionary. The dashed lines of the profile are nonevolutionary, and therefore by the criteria of this section cannot exist.

The point ($v_{x_1} = b_{x_1} = v_{x_2} = b_{x_2}$) does, however, correspond to an evolutionary discontinuity, but it is not a shock. It is called a "rotational" discontinuity by Landau and Lifshitz,[1] who show that in an incompressible fluid it is stable to small disturbances but that it has no steady-state structure. The latter statement means that in the presence of finite dissipation the rotational discontinuity eventually smooths out.

When $B_y$ vanishes, $c_f$ goes to $\max(b_x, a)$ and $c_s$ to the minimum. In this case, the shock (from Inequalities 3.103) is nonevolutionary if and only if the flow is super-Alfvénic ahead and sub-Alfvénic behind. This is also the condition for intermediate shocks.

The important special cases of "switch-on" and "switch-off" shocks can be discussed here too. For a "switch-on" shock, $B_{y_1} = 0$ and $v_{x_2} = b_{x_2}$. Hence, the coefficient $A_\ell$ in Equations 3.100 becomes unbounded, showing that the "switch-on" shock is unstable with respect to disintegration in the presence of normal Alfvén-wave disturbances. For a "switch-off" shock, $B_{y_2} = 0$ and $v_{x_1} = b_{x_1}$. Thus, the coefficient $A_\ell$ in Equation 3.99 becomes unbounded, indicating that "switch-off" shocks are likewise unstable. The conclusion that "switch-on" and "switch-off" shocks are not stable is an important one, particularly in view of a statement of Harold Grad that[36] "the 'switch-on' shocks must turn out to be stable or the whole structure of magnetohydrodynamic-shock theory is worthless. One must have 'switch-on' shocks to solve problems. For example, with the piston problem, one must have one to satisfy the boundary conditions."

## 4. Stability with Respect to Arbitrary Small Disturbances

In this section, the restriction imposed on the direction of the propagation vector of the incident disturbance wave is removed. Hence, the disturbance becomes three-dimensional, and the boundary conditions, Equations 3.22 through 3.29, are all coupled together. This means that an incident wave of any type may cause any of the seven types of diverging waves.

After we eliminate $\zeta_t$ and $\zeta_y$ from the eight boundary conditions, six equations remain from which to determine the amplitudes of diverging waves. Hence, as in the case of normal disturbances, evolutionary shocks occur when there are six waves diverging from the shock. By comparing Figures 3.6, 3.7, and 2.11, we can see that for normal disturbances the regions in which there are six diverging waves are those corresponding to $1 \to 2$, $2 \to 3$, and $3 \to 4$ shocks; and we must expect that when the incident propagation vector approaches the normal to the shock, the results of the previous section, without taking the splitting of boundary conditions into account, must be reproduced. The result of the splitting of boundary conditions is then to reduce even further the realm of evolutionary shocks by eliminating the $2 \to 3$ shock region. It appears, therefore, that the normal disturbance is the most severe case, and this is indeed true.

The problem of this section is as follows: For a given set of steady-state shock conditions, does the total number of diverging waves to be counted in the total disturbance vary with the angle of incidence of the incident wave? We shall prove that it does not; consequently, the results of the preceding section hold true for arbitrary angles of incidence. As in the previous paragraphs, the work here will be restricted, without loss of generality, to plane monochromatic waves. The work of this section is due to Kontorovich.[23]

To obtain the indicated result, we must first derive the general laws of reflection and refraction at a shock. Then, from these laws, we can determine the number and direction of diverging waves in any particular case. The complexity of the phase-velocity relation for magnetoacoustic waves (Equation 3.52), however, makes the analytical solution unwieldy; but fortunately there is a relatively simple graphical method, developed by Kontorovich, from which the desired results can be obtained in a very satisfactory manner. Then, as Kontorovich pointed out, the diverging waves must be classified as being on the left or right of the shock according to whether the group velocity points toward the left or right. By the use of a simple formula relating the group velocity to other parameters of the problem, we shall demonstrate graphically that the number of divergent waves which must be counted in the application of the boundary conditions is invariant with respect to the angle of incidence, thus proving the general validity of the results of Section 3.

General Laws of Reflection and Refraction at a Shock. These laws are derived from Equation 3.37, from the definition of phase velocity given by Equation 3.47, and from the following propositions:

I. The components in the plane of the shock of the propagation vectors of the incident and divergent waves are equal. (See page 37.)

II. The frequencies of the incident and divergent waves all measured in the same reference frame are equal.

If the phase velocity in the reference frame at rest with respect to the fluid is defined by $\omega_0 = \pm k v_{p_0}$, Equation 3.37 can be written

$$\omega - (kv) = \pm k v_{p_0} \tag{3.104}$$

in which $v_{p_0}$ is the absolute value of the phase velocity. The plus and minus signs are necessary for the following reason: In this problem, the incident-wave frequency is picked as a positive number in the reference frame at rest with respect to the shock. For some values of (kv), however, $\omega_0$ may be negative, in which case the negative sign in Equation 3.104 is chosen, because both k and $v_{p_0}$ are by definition positive. Negative $\omega_0$ means that the wave propagates in the direction $-\vec{k}$ in the reference frame at rest in the fluid, but still of course in the direction of $\vec{k}$ in the frame at rest with respect to the shock.

Look at Figure 3.9. As always, the plane of the shock is the y-z plane. In keeping with the choice made in setting up the boundary conditions, the propagation vector is assumed to lie in the x-y plane. In this reference frame, then, $\vec{B}$ must be allowed to have components in all three directions in general. The propagation vector is allowed to vary with respect to the shock and to $\vec{B}$ by varying the angles $\alpha$ and $\theta$. This may seem awkward at this point but is actually convenient.

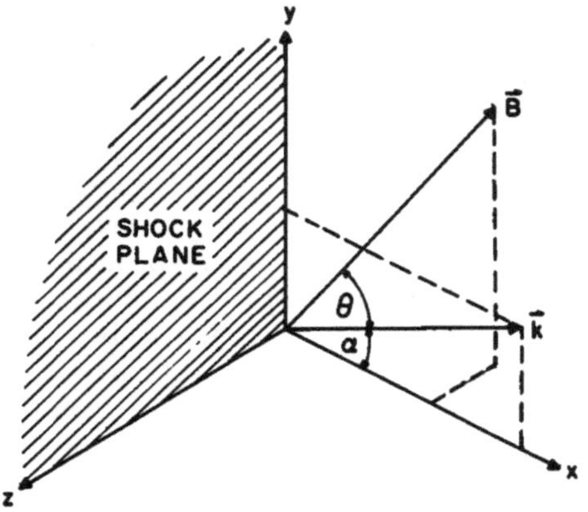


Figure 3.9. Orientation of the magnetic-field and propagation vectors with respect to the shock, which lies in the y-z plane

We obtain the equations for $v_{p_0}$ for the various waves from Equations 3.50, 3.51, and 3.52 by letting $v_i = 0$. It is evident that $v_{p_0}$ depends on the magnitudes a and b as well as on the angle $\theta$; however, for a given fluid state, $v_{p_0} = v_{p_0}(\theta)$, where $v_{p_0}(\theta)$ is a surface of revolution with the vector $\vec{B}$ as the axis. The x-y plane, in which by Proposition I all of the propagation vectors lie, cuts this surface of revolution. As $\alpha$ varies, the phase velocity of the wave corresponding to $\vec{k}$ varies along the curve of intersection of $v_{p_0}(\theta)$ and the x-y plane. Hence, it is equally valid to write $v_{p_0} = v_{p_0}(\alpha)$, where, from Figure 3.9, the relationship between $\alpha$ and $\theta$ is

$$\cos\theta = \frac{B_x}{B}\cos\alpha + \frac{B_y}{B}\sin\alpha \tag{3.105}$$

Now, we can write Equation 3.104 as

$$\omega - k_y v_y = k_x v_x \pm k\, v_{p_0}(\alpha)$$

Dividing by $k_y v_x$, we obtain

$$\frac{\omega - k_y v_y}{k_y v_x} = \frac{v_x \cos \alpha \pm v_{p_0}(\alpha)}{v_x \sin \alpha} \qquad (3.106)$$

The left side contains the quantities $\omega$ and $k_y$, which by Propositions I and II are the same for all waves, and also flow-velocity components fixed on each side of the shock for a given steady state. In fact, $\alpha$ is the only quantity in Equation 3.106 which varies among a given family of reflected waves derived from a given incident wave. Since $\alpha$ appears only on the right side, the law of reflection is that the left side of Equation 3.106 is the same for reflected and incident waves.

The right side of Equation 3.106 can be interpreted geometrically as cot $\psi$, where $\psi$ is defined as in Figure 3.10. The angle $\psi$ takes either of the two values shown, depending on the sign in front of $v_{p_0}$, but it is fixed for a given incident wave. If $\psi$ is defined by the relation

$$\cot \psi \equiv \frac{\omega - k_y v_y}{k_y v_x} \qquad (3.107)$$

The law of reflection becomes

$$\psi_{\text{incident}} = \psi_{\text{reflected}} \qquad (3.108)$$

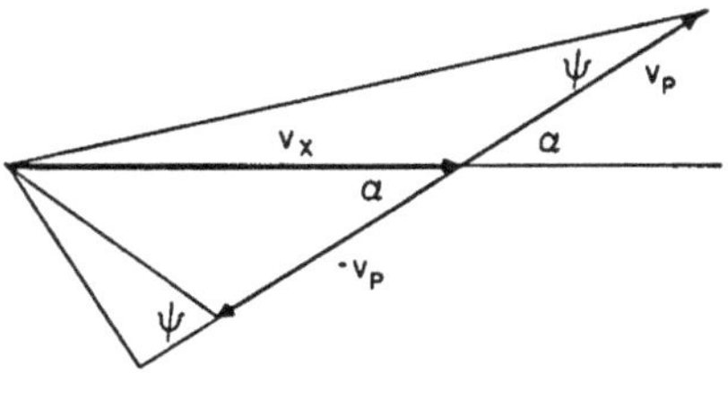


Figure 3.10. Construction of the invariant angle $\psi$

In terms of the geometry of Figure 3.10, the angle at the tip of the phase-velocity vector which subtends the vector component $v_x$ is either $\psi$ or $\pi - \psi$, depending on whether the sign in front of $v_{p_0}$ in Equation 3.106 is plus or minus for that particular wave. Hence, in terms of a construction, it is convenient to define an angle $\psi'$ as the angle from the tip of the phase-velocity vector subtending $v_x$. In terms of this angle, the law of reflection is

$$\psi_{(r)}' = \psi_{(i)}' \text{ or } \pi - \psi_{(i)}' \qquad (3.109)$$

The law of refraction is obtained by solving Equation 3.107 for $\omega/k_y$. Then, by Propositions I and II,

$$\frac{\omega}{k_y} = v_{x_1} \cot \psi_1 + v_{y_1} = v_{x_2} \cot \psi_2 + v_{y_2}$$

Hence,

$$\cot \psi_2 = \frac{v_{y_1} - v_{y_2} + v_{x_1} \cot \psi_1}{v_{x_2}} \tag{3.110}$$

Having obtained $\psi_2$, we can find the refracted waves.

Graphical Solution for the Diverging Waves. The graphical solution for the diverging waves can be obtained by plotting two loci. The first locus is the section of the surfaces $v_{p_0}(\theta)$ made by a plane parallel to the x-y plane and passing through the origin of $v_{p_0}(\theta)$. If $\vec{B}$ is in the x-y plane, these surfaces for fast- and slow sound waves and Alfvén waves are as shown in Figure 3.11 in a typical case. If $\vec{B}$ is out of the x-y plane, the new intersection can be obtained by rotating the $v_{p_0}(\theta)$ surfaces about the x-axis.

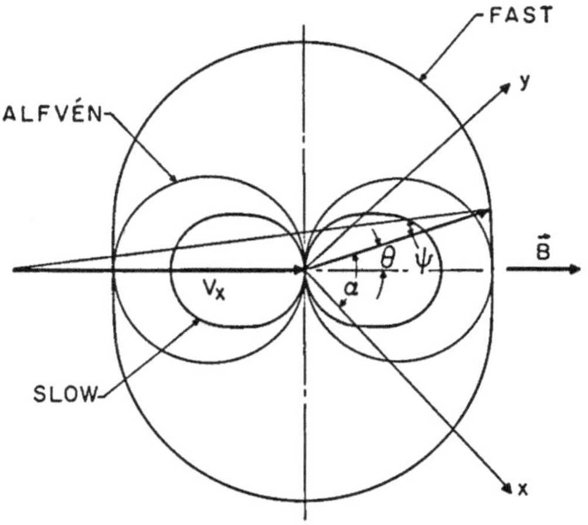


Figure 3.11. Phase-velocity diagram for fast-sound, slow-sound, and Alfvén waves

The second locus is the locus of all points which subtend $v_x$ at an angle of either $\psi$ or $\pi - \psi$ if the vector $v_x$ is placed in the position shown in Figure 3.11. The fact that this locus is a circle can easily be proved as follows: Consider the circle of radius r shown in Figure 3.12. Observe that

$$\tan \psi = \tan(\pi - \alpha - \beta) = \frac{\tan \alpha + \tan \beta}{\tan \alpha \tan \beta - 1}$$

But

$$\tan \alpha = \frac{y}{x + \sqrt{r^2 - a^2}}$$

and

$$\tan \beta = \frac{-y}{x - \sqrt{r^2 - a^2}}$$

Then, with the help of the equation of the circle,

$$x^2 + (y - a)^2 = r^2$$

it can easily be shown that

$$\tan \psi = \frac{\sqrt{r^2 - a^2}}{a} = \frac{\overline{AB}}{2a} \tag{3.111}$$

Since $\psi$ is independent of the coordinates of point P(x, y), it remains constant as P moves around the segment APB of the circle.

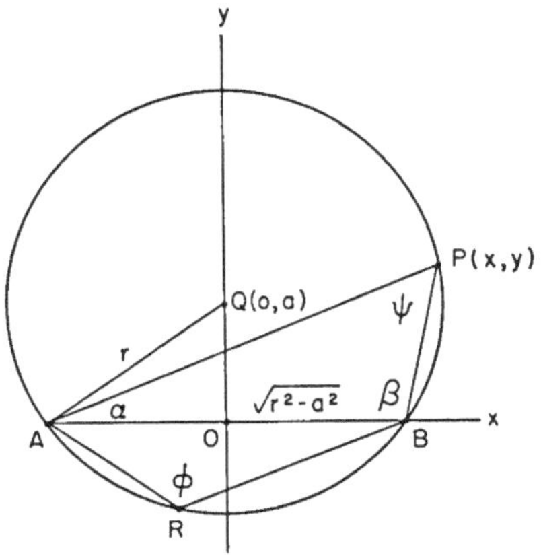


Figure 3.12. Locus of constant $\psi$

Consider the angle $\overline{ARB} = \phi$ in the lower portion of the circle of Figure 3.12. By rotating the circle 180 degrees around the line $\overline{AB}$, we can see that $\tan \phi$ is obtained from Equation 3.111 by replacing a by -a. Thus

$$\tan \phi = - \tan \psi$$

from which

$$\phi = \pi - \psi$$

The portion of the circle ARB is the locus of points which subtend $\overline{AB}$ at an angle $\pi - \psi$. Therefore, the entire circle is the locus desired.

We can now describe the graphical procedure. For a specific thermodynamic state and magnetic field, the $v_{p_0}(\theta)$ surfaces of revolution are defined. Then, for a specific value of $B_z/B$ (see Figure 3.9), the curves of intersection of $v_{p_0}(\theta)$ with the x-y plane are drawn, after which the length of $v_x$ as in Figure 3.11 is laid out and its perpendicular bisector is drawn. Next, a specific angle of incidence is chosen and a circle, centered on the perpendicular bisector, is drawn so that it passes through the end points of $v_x$ and the desired point on one of the phase-velocity curves. The intersections of this circle with the phase-velocity curves give the family of diverging waves corresponding to the given incident wave. For the refracted waves, a new angle must be calculated from Equation 3.110, and then the procedure is repeated.

Classification of Waves According to Group Velocity. Fortunately, it is not necessary to plot accurate diagrams such as those described above to obtain the results needed for the stability analysis. All that is required for this purpose is to ascertain, for a given steady-state condition, if the number of diverging waves can change when the angle of incidence changes. In this determination, one must recognize that the significant direction of a wave in determining whether it is reflected or refracted is the direction of energy propagation, i.e., the direction of the group velocity. For example, if the energy of a wave propagates toward the right, that wave must be counted as one of the waves making up the total disturbance on the right side of the shock even if its propagation vector points toward the left. (If one recalls that the component of the group velocity in the direction of the phase velocity is the phase velocity, it is easy to visualize how this can happen. See Figure 3.13.) With these considerations in mind, one can now study the three basic types of waves separately.

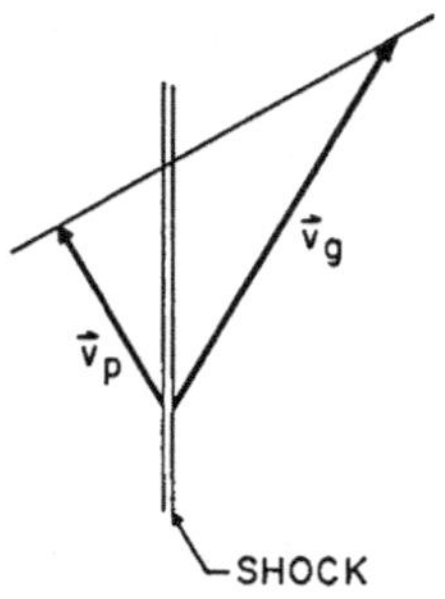


Figure 3.13. A possible relation between the group and phase velocities

a. Entropy wave: In this case, the wave energy propagates with the fluid. Hence, no matter how the shock is disturbed, there must always be one and only one entropy wave on the downstream side of the shock.

b. Alfvén waves: In this case, the energy propagates with the velocity given by Equation 3.55. Since this equation does not contain the propagation vector $\vec{k}$, the direction of energy propagation obviously cannot depend on the direction of $\vec{k}$. Thus, for this wave also, the same diverging waves result regardless of the angle of incidence.

c. Magnetoacoustic waves: As always, this is the only complicated case. Here the group velocity, given by Equation 3.56, does depend on $\vec{k}$. Nonetheless, the desired classification of waves is easy to accomplish by means of a simple relationship derived by Kontorovich. This relationship, however, has a derivation which is not at all obvious and would not likely be found by a direct search.

The derivation can be begun by combining Equations 3.106 and 3.107 to give

$$\cot \psi = \frac{v_x \cos \alpha \pm v_{p_0}(\alpha)}{v_x \sin \alpha} \tag{3.112}$$

Considering $\psi$ to be a function of $\alpha$ only, straightforward differentiation gives

$$\frac{\partial \psi}{\partial \alpha} = \frac{\sin^2 \psi}{v_x \sin^2 \alpha} \left[ v_x \pm \left( v_{p_0} \cos \alpha - \frac{\partial v_{p_0}}{\partial \alpha} \sin \alpha \right) \right] \tag{3.113}$$

Now consider Equations 3.56 and 3.57. If we substitute $\omega_0 = \pm k v_{p_0}$ (see Equation 3.104) and take into consideration Figure 3.9, they become

$$v_{g_i} = v_i \pm \frac{v_{p_0}{}^4 \dfrac{k_i}{k} - a^2 b^2 \cos \theta \dfrac{b_i}{b}}{v_{p_0} \left[ 2 v_{p_0}{}^2 - (a^2 + b^2) \right]} \tag{3.114}$$

$$v_{p_0}{}^2 \left[ 2 v_{p_0}{}^2 \quad (a^2 + b^2) \right] = v_{p_0}{}^4 - a^2 b^2 \cos^2 \theta \tag{3.115}$$

From Equation 3.115,

$$\frac{d v_{p_0}}{d \alpha} = - \left( \frac{a^2 b^2 \cos \theta}{v_{p_0}{}^4 - a^2 b^2 \cos^2 \theta} \right) v_{p_0} \frac{d \cos \theta}{d \alpha} \tag{3.116}$$

Now, using Equation 3.115 and Figure 3.9, we can write the x-component of Equation 3.114 as follows:

$$v_{g_x} = v_x \pm \left[ v_{p_0} \cos \alpha + \left( \frac{a^2 b^2 \cos \theta}{v_{p_0}{}^4 - a^2 b^2 \cos^2 \theta} \right) v_{p_0} \left( \cos \theta \cos \alpha - \frac{b_x}{b} \right) \right] \tag{3.117}$$

but from Equation 3.105,

$$\cos\theta\cos\alpha - \frac{b_x}{b} = \sin\alpha\,\frac{d\cos\theta}{d\alpha} \tag{3.118}$$

Substituting Equation 3.118 into Equation 3.117 and then comparing the result with Equation 3.116, we see that

$$v_{g_x} = v_x \pm \left( v_{p_0}\cos\alpha - \frac{dv_{p_0}}{d\alpha}\sin\alpha \right) \tag{3.119}$$

Substituting Equation 3.119 into Equation 3.113, we obtain the desired relation

$$\boxed{\frac{\partial\psi}{\partial\alpha} = \frac{\sin^2\psi}{v_x\sin^2\alpha}\,v_{g_x}} \tag{3.120}$$

Since $v_x$ can always be chosen greater than zero, Equation 3.120 shows that the normal component of the group velocity has the sign of $\partial\psi/\partial\alpha$. We shall now show that the sign of $\partial\psi/\partial\alpha$ can be visualized from the graphical solution; hence, the desired classification of waves can be made.

For illustrative purposes, a section of a typical $v_p(\theta)$ surface ($\chi = 3/2\sqrt{2}$) is shown in Figure 3.14. In this case the magnetic field lies in the x-y plane, i.e., the plane of the $\vec{k}$ vectors. The normal component of velocity, $v_x$, is chosen in this example to lie between the normal components of the fast and slow sound velocities, and, as indicated before, the centers of all possible $\psi$-circles lie on the perpendicular bisector of $v_x$. The $\psi$-circle, shown by a light solid curve, intersects the phase-velocity diagram at the points A, B, C, D. Points A and B are the fast and slow sound waves which move to the right with respect to the fluid and thus, because $v_x$ is positive, to

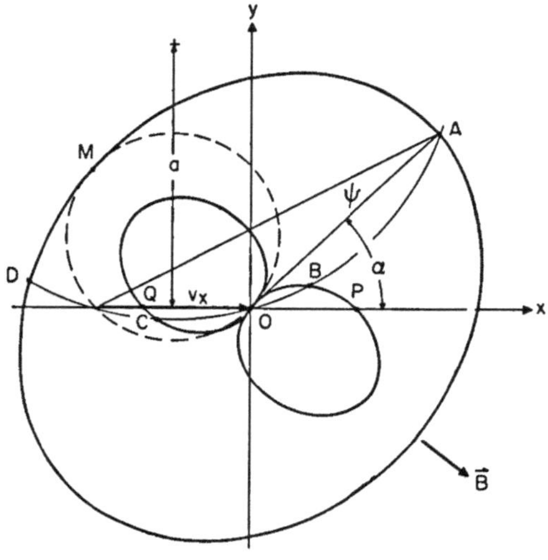


Figure 3.14. Intersection of the phase-velocity locus and the ψ-locus in a particular case

the right with respect to the shock. (Compare this with Figure 3.6.) Point C is on the lower side of $v_x$; it therefore subtends $v_x$ at an angle $\pi - \psi$. For such a point, $\omega_0 < 0$, and, as previously mentioned, the wave propagates in the direction $-\vec{k}$ with respect to the fluid. As a result, $\vec{k}$ must point from C to O, and the wave propagates to the right. This is the case of Figure 3.6, in which three magnetoacoustic waves move to the right. For point D, $\vec{k}$ goes from O to D; hence, this is a left-moving wave. Wave D could correspond to a fast wave incident on the shock from the right. Then A, B, and C would be the three waves reflected to the right. Or else, A (or B) could be the incident wave from the left, in which case D is the only wave reflected to the left. The waves B (or A) and C then would not appear because they can occur only on the right side of the shock.

Beginning with point A on the x-axis ($\alpha = 0$), consider the behavior of $\partial\psi/\partial\alpha$ for each wave separately, as $\alpha$ moves through all possible values. When $\alpha = 0$, the center of the $\psi$-circle has moved upward to infinity ($a \rightarrow +\infty$). As it comes down from infinity, point A moves upward, and $\alpha$ increases. Equation 3.111 shows that for fixed $v_x (= \overline{AB})$, decreasing $a$ means increasing $\psi$. Thus, as point A moves from the positive x-axis around to point M, $\partial\psi/\partial\alpha > 0$; so from Equation 3.120, $v_{g_x} > 0$. Hence, $\overline{OA}$ is a wave to be counted among diverging waves on the right side of the shock. Similarly, as point D moves from the negative x-axis up to M, $\alpha$ for the D-wave decreases and $\psi$ increases. Therefore, $\partial\psi/\partial\alpha < 0$, and $v_{g_x} < 0$. The D-wave in the indicated region is counted among those waves diverging to the left.

Suppose $a$ decreases further so that the $\psi$-circle no longer touches the fast-sound locus. The resulting behavior becomes clear if one realizes that Figure 3.12 is in reality a graphical solution of Equation 3.43 with 3.37 substituted. For a given state, and given values of $k_y$ and $\omega$, this is a general quartic for $k_x$. Thus, when the $\psi$-circle pulls away from the fast-sound curve, the solutions corresponding to A- and D-waves become complex conjugates. Then the exponential solution, Equation 3.31, has a positive and a negative real part. In order to satisfy the boundary condition of boundedness at infinity, only the negative real part can be kept on the right side, and only the positive real part on the left side. Thus, the A- and D-waves become surface waves which decay exponentially with the distance from the shock. They still must, however, be counted among waves which make up the total disturbance, but still it can be reckoned that the A-wave is on the right and the D-wave is on the left.

Let $a$ continue to decrease through zero and on to negative values. A value of $a$ will be reached for which the $\psi$-circle again intersects the fast locus. Then real A- and D-waves reappear, and, as $a \to \infty$, they come back to the positive and negative real axis, respectively. When $a$ is negative, it can be seen from Equation 3.111 that $\psi$ takes values between 90 and 180 degrees, and $\psi \to 180$ degrees as $a \to -\infty$. The angle of incidence $\alpha$ is still measured upwards from the positive x-axis, so $\alpha$ still increases as $\psi$ increases for the A-wave and decreases for the D-wave. As a result, $v_{g_x} > 0$ for the A-wave and $v_{g_x} < 0$ for the D-wave. Hence, regardless of the angle $\alpha$, the A-wave is on the right, and the D-wave is on the left.

Now, consider the slow-sound locus. As $a$ moves from $+\infty$ to $-\infty$, point B moves from point P along the locus for $y > 0$ to point O, then up and around to point Q. The angles $\psi$ and $\alpha$ both increase monotonically from 0 to $\pi$. Thus, $\partial\psi/\partial\alpha > 0$ always, and waves on this part of the slow-sound locus must lie on the right side of the shock. When B passes through the origin, $\omega_0$ passes through zero and becomes negative. When $\omega_0 = 0$, the disturbance is moving with the fluid, but it still must be reckoned as a legitimate wave propagating energy to the right, since, in this case, Figure 3.4 shows that the group velocity is directed along $\vec{B}$. Through the same range, point C moves from Q to P on the part of the locus for which $y < 0$, while $\alpha$ increases monotonically. Thus, again, $v_{g_x} > 0$ always, and this wave is to be counted on the right.

Suppose the tail of $\vec{v}_x$ extends to the left of the fast-sound locus. Then, the D-wave disappears for normal as well as all other angles of incidence, and the A-wave is always real. For the waves that remain, $\partial\psi/\partial\alpha$ has the same sign as before.

Suppose the tail of $\vec{v}_x$ lies to the right of the slow-sound curve. Then, the only change is in point C; it moves above the x-axis. Since the point Q is now to the left of the tail of $\vec{v}_x$, this wave moves to the left for normal incidence. As $a$ decreases from $+\infty$, point C moves upward, and $\alpha$ decreases from $\pi$. Therefore, $\partial\psi/\partial\alpha < 0$. A value of $a$ is eventually reached for which the $\psi$-circle is tangent to the slow-sound locus. Further decrease in $a$ causes the B- and C-waves to become complex; but, as described above, the reckoning of waves on the left and right sides does not change. Even further decrease in $a$ to negative values eventually produces real waves, but still the point C moves so that $\partial\psi/\partial\alpha < 0$. Hence, for this case also, the classification of diverging waves on the left and right is always the same as for normal incidence.

The case when $\vec{B}$ is not in the x-y plane remains to be considered. In reference to Figure 3.14, note that the shock lies in the plane $x = 0$ perpendicular to the paper. Keeping this plane perpendicular to the paper, rotate the phase-velocity surfaces of revolution, magnetic field and all, around the x-axis. This gives all the remaining freedom available to the plane of the $\vec{k}$ vectors. The most important thing to notice is that on the x-axis, the relative positions of $v_x$ and the intersections of the phase-velocity surfaces with the x-axis remain the same. The arguments on the sign of $\partial\psi/\partial\alpha$ depended only on this relationship; thus, again, this change in the angle of incidence does not affect the tally of left- and right-diverging waves.

This ends the proof that the conclusions obtained on stability with respect to normal small disturbances hold for arbitrary small disturbances.

## 5. The Fate of Nonevolutionary Shock Waves

The problem of the fate of nonevolutionary shocks has already been mentioned several times. In the present section we shall review briefly the existing literature on this subject because it gives further insight into the problem of shock stability.

General Discussion. In the literature mentioned above, this problem is usually enlarged to the consideration of the fate of discontinuities which at $t = 0$ have arbitrary parameters on the two sides unconnected by conservation laws. The present monograph, however, is more concerned with the behavior of shocks which seem to have some possibility of stability, and less with the detailed time history of initial arbitrary discontinuities. Therefore, this discussion focuses attention on intermediate shocks. In other words, the conservation laws and the requirement that entropy increase across a shock have separated from the domain of arbitrary discontinuities those which fall on the curves of Figure 2.11. Those which do not fall on these curves must certainly break up into a series of stable shocks, expansion waves, and possibly other types of discontinuities. The study of the influence of small disturbances showed by mathematical arguments that the intermediate shock region is nonevolutionary. It was further indicated that the nonevolutionary nature of intermediate shocks was solely due to their behavior in the presence of normal Alfvén disturbances — without which the shock would be evolutionary.

Additional insight into the problem came from a calculation of the amplitudes of normal Alfvén waves reflected and refracted from the shock. As the intermediate-shock region was approached, it was shown that the shock itself adds more and more energy to one of the diverging waves until, at the boundary, the wave amplitude becomes infinite. This analysis of the behavior of the shock was further strengthened by a mechanical analogy which gave some physical insight into the mechanism of instability.

Despite all this research, no direct study has yet been made to show what would happen to an intermediate shock if it were somehow formed. This type of insight can be obtained from the studies described below. As an illustration of the method of an analysis, we first make a study of the fate of a normal shock initially in the nonevolutionary region of Figure 3.8. Then we discuss the more general problem of nonevolutionary oblique shocks.

**Normal Shock Waves.** Consider a normal gas-dynamic shock in the presence of a normal magnetic field of magnitude such that the shock is in the nonevolutionary region of Figure 3.8. The analysis in Section 3 implies that a normal Alfvén wave will somehow destroy this steady-state configuration. Since all variables still depend only upon the coordinate normal to the shock, it is natural to postulate that the shock breaks up into a series of discontinuities still dependent only on the normal coordinate. Without solving a nonlinear unsteady-flow problem, it is not possible to follow the detailed behavior of the flow during the breakup phase; hence, progress can be made by assuming a new configuration and then exploring its consequences. If the new configuration is a sequence of stable discontinuities and is unique, one can be fairly sure that it represents the mode of breakup.

This problem is discussed by Liubarskii and Polovin,[24] who attribute the following solution (an unsatisfactory one) to Syrovatskii. Assume the configuration shown in Figure 3.15, consisting of two shocks separated by a region of fluid in which there is a tangential magnetic field. Clearly, this is a "switch-on," followed by a "switch-off," shock. Then, the conservation equations 2.1 through 2.6 give

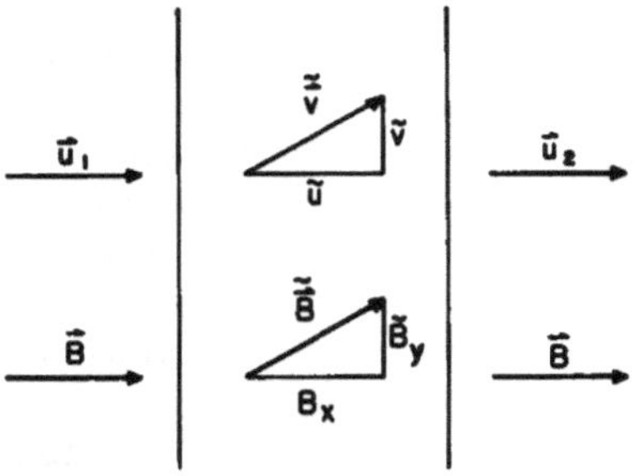


Figure 3.15. Postulated mode of breakup of a nonevolutionary normal shock

$$\rho_1 u_1 = \tilde{\rho}\tilde{u} = \rho_2 u_2 = G \qquad (3.121)$$

$$p_1 + \rho_1 u_1^2 = \tilde{p} + \tilde{\rho}\tilde{u}^2 + \frac{\tilde{B}_y^{\,2}}{2\mu} = p_2 + \rho_2 u_2^2 = F_x \qquad (3.122)$$

$$0 = \tilde{\rho}\tilde{u}\tilde{v} - \frac{B_x}{\mu}\tilde{B}_y = 0 \qquad (3.123)$$

$$h_1 + \frac{u_1^2}{2} = \tilde{h} + \frac{\tilde{u}^2 + \tilde{v}^2}{2} = h_2 + \frac{u_2^2}{2} = H \qquad (3.124)$$

$$0 = B_x\tilde{v} - \tilde{B}_y\tilde{u} = 0 \qquad (3.125)$$

in which the velocities are measured with respect to a reference frame which moves with the velocity of the original single shock. If $\tilde{B}_y \neq 0$, Equations 3.123 and 3.125 show that

$$\tilde{u}^2 = \frac{B_x^{\,2}}{\mu \tilde{\rho}} \tag{3.126}$$

Thus, the flow in the center region moves with the Alfvén velocity with respect to the reference frame at rest with respect to the original single shock. Recalling the properties of the "switch-on" and "switch-off" shocks, this result means that the two new shocks move with the same velocity as the original one.

Now assume the perfect-gas relations

$$p = \frac{\rho a^2}{\gamma}, \qquad h = \frac{a^2}{\gamma - 1}$$

and eliminate $\tilde{u}^2$ and $\tilde{v}^2$ from Equations 3.122 and 3.124. Then, these two equations become

$$\frac{\tilde{\rho}\tilde{a}^2}{\gamma} + \frac{B_x^{\,2}}{\mu} + \frac{\tilde{B}_y^{\,2}}{2\mu} = F_x$$

$$\frac{\tilde{a}^2}{\gamma - 1} + \frac{B_x^{\,2}}{2\mu\tilde{\rho}} + \frac{\tilde{B}_y^{\,2}}{2\mu\tilde{\rho}} = H$$

If $\tilde{a}^2$ is eliminated between these two equations, and the substitution $\tilde{\rho} = G^2\mu/B_x^{\,2}$ (from Equations 3.121 and 3.126) is made,

$$\frac{\tilde{B}_y^{\,2}}{2\mu} = \gamma F_x - (\gamma + 1)\frac{B_x^{\,2}}{2\mu} - (\gamma - 1)\frac{G^2\mu H}{B_x^{\,2}}$$

Substituting for $F_x$, $H$, and $G^2$ quantities on either side of the original shock, one can manipulate the result into the form

$$\frac{\tilde{B}_y^{\,2}}{2\mu} = \frac{\rho(u^2 - b_x^{\,2})^2}{b_x^{\,2}}\left[f(B_x) - \frac{\gamma + 1}{2}\right] \tag{3.127}$$

in which

$$f(B_x) = \frac{u^2 - a^2}{u^2 - b_x^{\,2}} \tag{3.128}$$

The possibility of the steady-state configuration of Figure 3.15 rests upon whether $f_{1,2}(B_x) \gtrless \frac{\gamma + 1}{2}$, that is, on whether $\widetilde{B}_y^2$ is real or imaginary. The behavior of $f_1$ and $f_2$ as functions of $B_x$ is shown in Figure 3.16 by superimposing plots of these functions on Figure 3.8. Outside the region $\overline{PQ}$ on the line

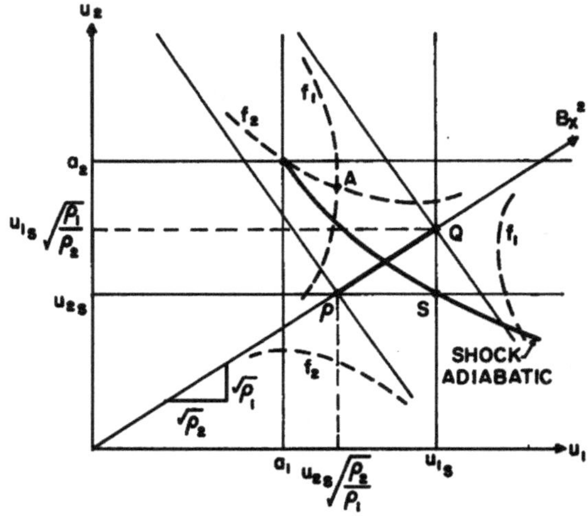


Figure 3.16. Regions in which configuration of Figure 3.15 is possible

$u_2/u_1 = \sqrt{\rho_1/\rho_2}$, either $f_1$ or $f_2$ is negative; hence, $\widetilde{B}_y^2 < 0$, and the shock does not split. Inside the region $\overline{PQ}$, certainly $f_1(B_x) > (\gamma + 1)/2$ near point Q, and $f_2(B_x) > (\gamma + 1)/2$ near point P. The only question is the relative magnitude of $f_1$ ($= f_2$) compared with $(\gamma + 1)/2$ at point A.

From the equality $f_1 = f_2$ the value of $B_x^2/\mu$ at point A can be found and expressed in the form

$$\frac{B_x^2}{\mu\rho_1} = b_{x_1}^2 = \frac{u_2^2 a_1^2 - u_1^2 a_2^2}{u_2^2 - a_2^2 - \frac{\rho_1}{\rho_2}(u_1^2 \quad a_1^2)}$$

After this expression is substituted into Equation 3.128 for $f_1$, the factor $u_1^2 - a_1^2$ cancels out. Then, with the help of the continuity equation, $f_1(A)$ reduces to

$$f_1(A) = \frac{u_2(u_2 - u_1) - \frac{1}{\rho_2}(\rho_2 a_2^2 - \rho_1 a_1^2)}{u_2(u_2 - u_1)}$$

But from Equations 3.121, 3.122, and the expression $p = \rho a^2/\gamma$, $\rho_1 a_1^2 - \rho_2 a_2^2 = \gamma \rho_2 u_2 (u_2 - u_1)$. Hence

$$f_1(A) = \gamma + 1 > \frac{\gamma + 1}{2}$$

which implies that both $f_1$ and $f_2$ are greater than $(\gamma + 1)/2$ everywhere within the interval $\overline{PQ}$. Hence, everywhere in $\overline{PQ}$, $\tilde{B}_y^2 > 0$ and the configuration of Figure 3.14 is possible.

This is an example – probably the simplest – of the type of algebraic analysis used to find possible stable configurations to replace a single nonevolutionary shock. In the present case, however, the new configuration is a "switch-on" shock followed by a "switch-off" shock; and it was previously shown that these shocks must also be classed as nonevolutionary. There are no other types of discontinuities in magnetohydrodynamics in which $B_y$ is zero on one side and finite on the other. Hence, the normal nonevolutionary shock cannot break up into any of the known types of magnetohydrodynamic discontinuities. Moreover, it cannot break up into a series of weaker normal shocks, because it is well known that the rear shock overtakes the forward one. Syrovatskii[35] has concluded that it "is converted into a certain nonstationary flow under the action of infinitesimal perturbations" and also that "this nonstationary flow can evidently contain one or several diverging shock waves." At the present time this seems to be about all that can be said.

Oblique Shock Waves. Liubarskii and Polovin[24] discuss the above problem as the zeroth-order approximation to the problem of oblique shocks in which $B_y$ is small. Unfortunately, they do not consider the case $B_y = 0$ as an important special case in itself, but go on to the study of oblique shocks in which only second-order terms in $B_y$ are dropped. (In the zeroth-order approximation first-order terms in $B_y$ are dropped.) Within the restriction that the shock is weak, they conclude that nonevolutionary shocks break up into a sequence of discontinuities, which they describe very briefly.

The same problem has been solved by Gogosov in a series of three papers[20,21,22] without any restrictions on the magnitude of the transverse magnetic field or on the strength of the shock. In this generality, analytic solution is not possible; therefore Gogosov resorted to a numerical-graphical procedure in which the main difficulty is in organizing and cataloguing the results. Gogosov reasoned that from the similarity properties of the problem, the nonevolutionary shock must break up into some combination of fast shocks and/or expansion waves, Alfvén (or rotational) discontinuities, and slow shocks and/or expansion waves going in both directions and separated by a contact discontinuity. He has stated that altogether there are 648 different possible combinations of waves and discontinuities which may be realized, depending on the parameters ahead of and behind the shock.

Of course, if attention were restricted to intermediate shocks, this number would have to be considerably reduced. Unfortunately, his results are presented in a form in which it is not at all obvious how to tell which of his regions or curves refer to intermediate shocks. Contrary to the case of normal nonevolutionary shocks, it is easy, however, to visualize possible modes of breakup of oblique intermediate shocks, because there is a tangential magnetic field on both sides.

One further point should be made. A requirement for a particular mode of breakup of a nonevolutionary shock to be possible is that the separate discontinuities into which it is resolved be stable. This is true of the discontinuities used by Gogosov if the fluid is assumed perfect; however, Alfvén and contact discontinuities diffuse in a real gas. Therefore, the solutions of Gogosov can be considered as correct only for times that are small compared with the diffusion times. After a longer period has elapsed, the flow must take some more general nonstationary form.

In conclusion, it appears that the apparently simple case of a normal nonevolutionary shock is the only one for which a specific mode of breakup cannot be suggested; thus, solution of the nonlinear unsteady-flow problem for this case may be both interesting and profitable.

Chapter 4

# EQUATIONS OF THE STEADY-STATE SHOCK LAYER

Up to this point we have assumed that the shock is a discontinuity in a perfect fluid. On this basis we have indicated that only fast and slow shocks satisfy necessary conditions for stability. Intermediate shocks and the two borderline cases — "switch-on" shocks on the fast-shock side and "switch-off" shocks on the slow-shock side — apparently could not be formed in nature because any disturbance containing a normal Alfvén wave as one component must drain away the entire energy of the shock into the diverging waves. The arguments from which this conclusion was derived are uncomfortably simple, and of course take no account whatever of the dissipative mechanisms which form the steady-state shock in the first place.

The ultimate problem which, it seems, must be solved in order to understand shock stability completely is the unsteady nonlinear problem of the behavior of a nonevolutionary shock in the presence of dissipative effects. Before tackling such a problem — if it can ever be done mathematically — one must establish the existence of the steady-state shock layer and understand its structure. Solution of the latter problem is important also because it will provide a detailed mathematical theory to use for comparison with experiment. The remainder of this monograph concerns this problem of existence and the qualitative behavior of the shock layer.

The problem of shock structure in gas dynamics has a long history, reviewed and extended, for example, by Gilbarg and Paolucci[11] and by Grad[37]. A long debate ensued over the appropriate theory upon which to base detailed shock-layer studies. Since the thickness of the shock layer is of the order of a mean free path, no one denied the appropriateness of a statistical treatment of the particle interactions if an exact mathematical solution were possible; however, the need to resort to approximations and the inability to prove convergence of those approximations removed the possibility of a theoretical settlement of this question.

Recent experiments, reported by Talbot and Sherman,[38] on the structure of shock waves in argon show that the profile of weak shocks agrees with solutions based on the Navier-Stokes and Burnett equations, but not so well with the Grad 13-moment equations. More recently, Ziering, Ek, and Koch[39] have concluded that the Navier-Stokes solution is best for weak and moderate-strength shocks and that the Mott-Smith solution is best for strong shocks. In another approach to the problem, Haviland[40] has obtained substantially the same results by means of computer studies based on a Monte-Carlo method.

For ionized gases in the presence of magnetic fields, the literature on shock structure falls into two groups. The first group deals with highly rarefied plasma and the possibility of finding shocks of width many times as small as a mean free path, since there is some experimental evidence that such shocks exist.[41] In this group, referred to as "collisionless-shocks," the research is summarized, for example, in a paper by Morawetz.[42]

The second group of papers follows a line of attack to be extended here. These studies consider the gas as a single fluid, and take the Navier-Stokes approximation for the pressure tensor and heat-flux vector. Use of the Navier-Stokes approximation implies that the stress tensor is isotropic, and hence that the collision frequency is much greater than the cyclotron frequency. The first paper in this group was contributed by Marshall,[43] who treated the case in which the magnetic field is parallel to the plane of the shock ($B_n = 0$). Then Ludford[44] set up the general case of an arbitrarily directed magnetic field, but restricted his detailed discussion to the case $B_n = 0$ and to "switch-on" and "switch-off" shocks. Moreover, he neglected thermal conductivity and shear viscosity.

Germain[13] has discussed the existence of oblique magnetohydrodynamic shocks in the most detail. He proves, without neglecting any of the dissipation coefficients, and for an arbitrary real gas, that fast shocks always exist and that intermediate shocks cannot be considered in the limit as discontinuities in a perfect fluid if the dissipation coefficients are allowed to vanish in an arbitrary order. The treatment of slow shocks is more complicated because of a mathematical difficulty; but for a perfect gas with negligible shear viscosity and thermal conductivity, he seems to have found cases in which slow shocks cannot exist. In a later paper, Kulikovskii and Liubimov[45] reworked the same problem; however, because they unfortunately failed to include the bulk-viscosity term in their energy equation, their results are open to question.

The main criticism of all of these papers in the second group is that the macroscopic formulation of magnetohydrodynamics, upon which they are based, is not valid for shocks. These papers are, nevertheless, important steppingstones to a more accurate formulation of the problem. When changes occur over lengths of the order of a mean free path, we show in the present study that the collision term in the generalized Ohm's law is of the order of the inertia terms, and that inside the shock the electrical forces can be as strong as the magnetic forces. Addition of the current-inertia effect raises the order of the system by one and permits the possibility of complex eigenvalues in the solution linearized near singular points. The electrical body force caused by charge separation — to be expected because of the greater mobility of the electrons — produces no modification in the existence proofs but significantly alters the structure of the shock.

The purpose of the remainder of this monograph is then to determine what shocks can exist and to find their qualitative behavior within the framework of a more realistic set of equations for the steady-state one-dimensional shock layer. These will be continuum equations, and the Navier-Stokes approximation will be used. But even then there are two possible approaches.

The first is the two-fluid approach in which one fluid is the ions and one the electrons. Then, Navier-Stokes equations can be found for each fluid by computing moments of the Boltzmann equation in a well-known manner. In these equations there are collision terms in both the momentum and energy equations as well as a set of dissipation coefficients in both. All in all, this formulation results in eight separate dissipation coefficients, some of which may of course be negligible. It does, however, have the added freedom that the downstream electron and ion temperatures need not be equal.

The second and preferred approach is the single-fluid approach, the most rigorous formulation of which has been recently given by H. S. Green.[46] Here the continuity, momentum, and energy equations are the usual continuum magnetohydrodynamic equations with the electrical body force included in the momentum equation. The main difference is that the Ohm's law is obtained as the difference between the momentum equations for ions and the one for electrons, and it is not simplified as is possible in macroscopic nonrelativistic flow problems. Although the second approach results in one less degree of freedom, it is far simpler and can also be more directly correlated with previous work on shocks in the continuum region. It seems to be the logical next step to take in the study of collision-dominated magnetohydrodynamic shocks.

In this chapter we derive the steady-state shock-layer equations. They are based primarily on the formulation of the macroscopic equations as given by H. S. Green[46]; however, the works of Spitzer,[47] Delcroix,[48] Hirshfelder, Curtiss, and Bird,[49] de Groot,[50] and others have proved invaluable in arriving at the final formulation. We must also mention that the whole motivation for this study came from an analysis of the generalized Ohm's law given by Professor Stanislaw Olbert, of M. I. T. His analysis is presented in the class notes for the course on Cosmic

Physics taught in the spring of 1961 at M.I.T. by Professors Bruno Rossi and Olbert.

## 1. The Basic Equations of Magnetohydrodynamics

The continuum equations of magnetodydrodynamics given in this section are valid in an intermediate range of temperatures, high enough so that quantum mechanical effects are unimportant but low enough so that relativistic effects, radiation pressure, and energy density may be neglected.* As is well known, this intermediate range is extremely broad and covers many practical applications. Furthermore, it is assumed in the derivation that changes in all macroscopic quantities are small in distances comparable to the distance across which particles are correlated.[46] This assumption is also implicit in the Navier-Stokes equations and accounts for the fact that they produce correct quantitative results only for weak shocks in ordinary gas dynamics. In a plasma, the long-range Coulomb forces extend the correlation distance much further than in a neutral gas; but then the interaction between particles — i.e., collisions — also occurs over much greater distances. Thus, there is good reason to believe that plasma equations based on the Navier-Stokes approximation will be quantitatively valid for weak magnetohydrodynamic shocks too. The extent to which this is correct can only be found after the theory has been worked out in detail.

The Continuity, Momentum, and Energy Equations. The derivation of the macroscopic equations of magnetohydrodynamics is fundamentally based upon the Liouville equation, which states that the volume in phase space occupied by a given group of particles is constant. First, the forces which appear in the Liouville equation are separated into those due to the electromagnetic field existing within the plasma and those ostensibly due to collisions. Next, the equation is multiplied by a function of velocity only, $\theta(\vec{v})$, and integrated over velocity space. Then, letting $\theta(\vec{v})$ equal $m_i$, $m_i v_i$, $\frac{1}{2} m_i v_i^2$, one finds, respectively, the equations of continuity, momentum, and energy for the $i^{th}$ species of particles within the plasma. Summing over all species, one obtains the equations of continuity, momentum, and energy of the entire plasma, which are

$$\frac{\partial \rho}{\partial t} + \frac{\partial \rho u_k}{\partial x_k} = 0 \qquad (4.1)$$

---

* The effect of radiation energy flux is taken into account in that it merely increases the coefficient of thermal conductivity.

$$\frac{\partial \rho u_i}{\partial t} + \frac{\partial}{\partial x_k}(P_{ik} + \rho u_i u_k) = \rho_c E_i + (\vec{j} \times \vec{B})_i \tag{4.2}$$

$$\rho \frac{dU}{dt} + P_{ik}\frac{\partial u_i}{\partial x_k} + \frac{\partial Q_k}{\partial x_k} + u_i\left[\rho_c E_i + (\vec{j} \times \vec{B})_i\right] = j_i E_i \tag{4.3}$$

in which $u_i$ is the bulk velocity of the fluid, $P_{ik}$ is the stress tensor, $Q_k$ is the heat flux, $U$ is the internal energy of the fluid per unit mass, $\rho_c$ is the total charge density, and $\vec{j}$ is the total current.

These equations are in the form presented by Green[46] (converted to MKS units); however, for steady-state problems, it is preferable to express them all in the standard form for conservation laws, i.e., in a form which explicitly states that the time rate of change of the conserved density plus the divergence of its flux is equal to zero. This is easily done with the help of Maxwell's equations, which, in MKS units, can be written in the following tensor form:

$$\epsilon_{ijk}\frac{\partial\left(\frac{B_k}{\mu}\right)}{\partial x_j} = \frac{\partial \epsilon E_i}{\partial t} + j_i \tag{4.4}$$

$$\epsilon_{ijk}\frac{\partial E_k}{\partial x_j} = -\frac{\partial B_i}{\partial t} \tag{4.5}$$

$$\frac{\partial \epsilon E_i}{\partial x_i} = \rho_c \tag{4.6}$$

$$\frac{\partial B_i}{\partial x_i} = 0 \tag{4.7}$$

As pointed out by Green, for analysis of plasmas these equations should be — and have been — expressed in the form in which $\vec{j}$ and $\rho_c$ represent the <u>total</u> current and charge densities, respectively. Then $\epsilon$ and $\mu$ have their free-space values.

With the help of Equations 4.4 through 4.7, the Lorentz force in Equation 4.2 can be expressed in the form

$$\rho_c E_i + (\vec{j} \times \vec{B})_i = \frac{\partial}{\partial x_k}\left[\epsilon E_i E_k + \frac{B_i B_k}{\mu} - \frac{1}{2}\left(\epsilon E^2 + \frac{B^2}{\mu}\right)\delta_{ik}\right]$$

$$- \frac{\partial}{\partial t}(\epsilon \vec{E} \times \vec{B})_i \qquad (4.8)$$

The derivation of Equation 4.8 is straightforward and can be found, for example, in Reference 51. Substitution of Equation 4.8 into 4.2 gives

$$\frac{\partial}{\partial t}\left[\rho u_i + (\epsilon \vec{E} \times \vec{B})_i\right]$$

$$+ \frac{\partial}{\partial x_k}\left[P_{ik} + \rho u_i u_k - \epsilon E_i E_k - \frac{B_i B_k}{\mu} + \frac{1}{2}\left(\epsilon E^2 + \frac{B^2}{\mu}\right)\delta_{ik}\right] = 0$$

$$(4.9)$$

which is the desired form of the momentum equation.

Using Equations 4.1 and 4.2, we can convert Equation 4.3 into the form

$$\frac{\partial}{\partial t}\left(\rho U + \rho\frac{u^2}{2}\right) + \frac{\partial}{\partial x_k}\left[\rho u_k\left(U + \frac{u^2}{2}\right) + P_{ik}u_i + Q_k\right] = j_k E_k$$

$$(4.10)$$

Then, using Equations 4.4 and 4.5,

$$j_k E_k = -\frac{\partial}{\partial x_k}\left(\vec{E} \times \frac{\vec{B}}{\mu}\right)_k - \frac{\partial}{\partial t}\left(\frac{\epsilon E^2}{2} + \frac{B^2}{2\mu}\right)$$

Hence, Equation 4.10 becomes

$$\frac{\partial}{\partial t}\left[\rho U + \rho\frac{u^2}{2} + \frac{\epsilon E^2}{2} + \frac{B^2}{2\mu}\right]$$

$$+ \frac{\partial}{\partial x_k}\left[\rho u_k\left(U + \frac{u^2}{2}\right) + P_{ik}u_i + Q_k + \left(\vec{E} \times \frac{\vec{B}}{\mu}\right)_k\right] = 0$$

$$(4.11)$$

which is the desired form of the energy equation.

The Pressure Tensor. In the Navier-Stokes approximation, the stress tensor is

$$P_{ik} = p\,\delta_{ik} - \eta\left(\frac{\partial u_i}{\partial x_k} + \frac{\partial u_k}{\partial x_i} - \frac{2}{3}\frac{\partial u_\ell}{\partial x_\ell}\,\delta_{ik}\right) - \zeta\,\frac{\partial u_\ell}{\partial x_\ell}\,\delta_{ik} \qquad (4.12)$$

in which $p$ is the scalar pressure and it can be shown[18] that the two viscosity coefficients $\eta$ and $\zeta$ are positive definite. By expressing the stress tensor in the above form, we again emphasize that the collision frequency must be much greater than the cyclotron frequency.

The Heat-Flux Vector. For the case of a one-component fluid, the heat-flux vector $\vec{Q}$ is assumed in the Navier-Stokes approximation to be simply proportional to the temperature gradient. In a fluid in which there is more than one component — for example, ions and electrons in a plasma — it has been shown both by the theory of irreversible processes in thermodynamics[50] and by kinetic theory that the temperature gradient is only one of several causes of $\vec{Q}$.

The basic kinetic-theory definition of $\vec{Q}$ is

$$\vec{Q} = \sum_j \int \frac{1}{2} m_j w_j^2 \vec{w}_j f_j(\vec{r}, \vec{v}_j, t)\, d\vec{v}_j$$

in which $\vec{w}_j$ is the diffusion velocity $\vec{v}_j - (\vec{v}_j)_{av}$ of a particle of the $j^{th}$ species, and $f_j$ is the corresponding distribution function. Thus $\vec{Q}$ is the total flux of random kinetic energy due to any cause whatever, only one of which is a temperature gradient. As is well known from kinetic theory, the above sequence of magnetohydrodynamic equations can be closed at the Navier-Stokes level of approximation only by expressing $P_{ik}$ and $\vec{Q}$ in terms of lower-order dependent variables by means of a phenomenological argument. This was done for $P_{ik}$ in Equation 4.12. For $\vec{Q}$, the proper phenomenological expression is not so well accepted.

De Groot[50] represents the heat flux in the form

$$\vec{Q} = \sum_{k=1}^{n} L_{uk}\vec{X}_k + L_{uu}\vec{X}_u$$

in which

$$\vec{X}_k = \vec{F}_k - T\,\mathrm{grad}\left(\frac{\mu_k}{T}\right)$$

$$\vec{X}_u = - \frac{\text{grad } T}{T}$$

and $L_{uk}$, $L_{uu}$ are phenomenological coefficients deducible from kinetic theory; $\vec{F}_k$ is the external body force on the $k^{th}$ species — in the present case the Lorentz force — and $\mu_k$ is the chemical potential of the $k^{th}$ species. This is also the form accepted by Green.[46]

The use of the chemical potential in the present problem is inconvenient, but fortunately there is another equally valid formulation which proves to be easier to use. Hirshfelder, Curtiss, and Bird[49] have derived the following expression for heat flux:

$$\vec{Q} = - \lambda \frac{\partial T}{\partial \vec{r}} + \frac{5}{2} kT \sum_i n_i (\vec{w}_i)_{av} + \frac{kT}{n} \sum_{i,j} \frac{n_j D_i^T}{m_i \mathscr{D}_{ij}(1)} \left[ (\vec{w}_i)_{av} - (\vec{w}_j)_{av} \right] \tag{4.13}$$

This is their Equation 7.4-64. The first term is the usual temperature gradient, in which T is the kinetic-theory temperature defined by

$$\frac{3}{2} kT = \frac{\sum_i n_i \dfrac{m_i (w_i)^2_{av}}{2}}{\sum_i n_i}$$

The second term is the direct energy diffusion of each type of particle with respect to the mean motion of the fluid, and the third term is a more complex thermal diffusion effect. Equation 4.13 is based on the usual Boltzmann equation in which just binary collisions are taken into account, but the particular force law enters only in the exact form of the two diffusion coefficients $D_i^T$ and $\mathscr{D}_{ij}(1)$.

Consider a fully ionized plasma, and for simplicity let it be singly ionized. Then from either Reference 46, 47, or 48, the bulk velocity $\vec{u}$ and total current $\vec{j}$ are given by

$$\rho \vec{u} = m_+ n_+ \vec{u}_+ + m_- n_- \vec{u}_- \tag{4.14}$$

$$\vec{j} = e(n_+ \vec{u}_+ - n_- \vec{u}_-) \tag{4.15}$$

in which

$$\rho = m_+ n_+ + m_- n_- \tag{4.16}$$

It is convenient to write the total current $\vec{j}$ as the sum of the conduction current $\vec{J}$ and the convection current $\rho_c \vec{u}$. Thus,

$$\vec{j} = \vec{J} + \rho_c \vec{u} \tag{4.17}$$

in which the charge density $\rho_c$ is

$$\rho_c = e(n_+ - n_-) \tag{4.18}$$

Equation 4.17 is the defining equation for $\vec{J}$. Now referring to Equation 4.13,

$$\sum_i n_i (\vec{w}_i)_{av} = n_+(\vec{u}_+ - \vec{u}) + n_-(\vec{u}_- - \vec{u}) \tag{4.19}$$

$$(\vec{w}_+)_{av} \quad (\vec{w}_-)_{av} = \vec{u}_+ - \vec{u}_- \tag{4.20}$$

and the solution of Equations 4.14 and 4.15 for $\vec{u}_+$ and $\vec{u}_-$ is

$$\vec{u}_+ = \frac{1}{n_+(1 + m_+/m_-)} \left( \frac{\rho \vec{u}}{m_-} + \frac{\vec{j}}{e} \right) \tag{4.21}$$

$$\vec{u}_- = \frac{1}{n_-(1 + m_-/m_+)} \left( \frac{\rho \vec{u}}{m_+} - \frac{\vec{j}}{e} \right) \tag{4.22}$$

Using Equations 4.21, 4.22, and 4.16, Equation 4.19 can be reduced to the form

$$\sum_i n_i (\vec{w}_i)_{av} = \left( \frac{m_+ - m_-}{m_+ + m_-} \right) \left[ (n_+ - n_-)\vec{u} - \frac{\vec{j}}{e} \right]$$

But Equations 4.17 and 4.18 allow this expression to be written

$$\sum_i n_i (\vec{w}_i)_{av} = - \left( \frac{m_+ \quad m_-}{m_+ + m_-} \right) \frac{\vec{J}}{e} \approx - \frac{\vec{J}}{e} \tag{4.23}$$

since $m_-/m_+ \ll 1$.

In a similar manner, Equation 4.20 can be reduced to

$$\vec{u}_+ - \vec{u}_- = \left[ \frac{m_+ n_+ + m_- n_-}{n_+ n_- (m_+ + m_-)} \right] \frac{\vec{J}}{e} \approx \frac{\vec{J}}{en_-} \tag{4.24}$$

if $n_+$ and $n_-$ do not differ greatly from each other.

Substituting Equation 4.23 and 4.24 into Equation 4.13,

$$\vec{Q} = -\kappa \frac{\partial T}{\partial \vec{r}} - \kappa' \vec{J} \tag{4.25}$$

in which $\kappa = \lambda$ is used to conform to the notation used in the following chapter, and

$$\kappa' = \frac{kT}{e} \left[ \frac{5}{2} - \frac{D_+^T}{nm_+ \mathscr{D}_{+-}(1)} + \frac{D_-^T}{nm_- \mathscr{D}_{-+}(1)} \right] \tag{4.26}$$

For the present study, $\kappa'$ will be considered as an additional phenomenological coefficient, calculable from kinetic theory.

Equation 4.25 will be used to represent the heat-flux vector in this work, since it is simpler than the form given by de Groot and is equally justified on a phenomenological basis, the difference being that de Groot represents the cross-coupling effect by means of forces, whereas in Equation 4.25 it is represented by the resulting effect.

The Generalized Ohm's Law. If the momentum equation for the $a^{th}$ species is multiplied by $e_a/m_a$ — the charge to mass ratio — and the resulting equation is summed over all species, the generalized Ohm's law is obtained. Green[46] gives this equation in the following form (his Equation 52):

$$\frac{d}{dt}(\vec{j} - \rho_c \vec{u}) + (\vec{j} - \rho_c \vec{u}) \cdot \nabla \vec{u} + (\vec{j} - \rho_c \vec{u}) \nabla \cdot \vec{u}$$
$$+ \sum_a \left( \frac{e_a}{m_a} - \frac{\rho_c}{\rho} \right) \left( \operatorname{div} \vec{\vec{P}}_a - \rho_{c_a} \vec{E} - \vec{j}_a \times \vec{B} \right)$$
$$= \sum_{a,b} \frac{e_a^2 e_b}{m_a} \int \nu_{ab}^{(-)} \frac{\vec{r}}{r^3} \, dr^3 \tag{4.27}$$

in which $d/dt = (\partial/\partial t) + \vec{u} \cdot \nabla$ and the collision integral on the right side represents the interaction between ions and electrons.

Since $m_-/m_+ \ll 1$,

$$\frac{e_+}{m_+} - \frac{\rho_c}{\rho} \approx \frac{en_-}{m_+ n_+} \qquad \frac{e_-}{m_-} - \frac{\rho_c}{\rho} \approx -\frac{e}{m_-}$$

Then from Equations 4.15 and 4.18, $\vec{j}_a = e_a n_a \vec{u}_a$ and $\rho_{c_a} = e_a n_a$. Consequently, with the help of Equation 4.17, Equation 4.27 can be written in tensor notation as follows:

$$\frac{\partial J_i}{\partial t} + u_k \frac{\partial J_i}{\partial x_k} + J_k \frac{\partial u_i}{\partial x_k} + J_i \frac{\partial u_k}{\partial x_k} + \frac{en_-}{m_+ n_+} \frac{\partial P_{+ik}}{\partial x_k} - \frac{e}{m_-} \frac{\partial P_{-ik}}{\partial x_k}$$

$$- \frac{e^2 n_-}{m_-} E_i - e^2 n_- \left[\left(\frac{\vec{u}_+}{m_+} + \frac{\vec{u}_-}{m_-}\right) \times \vec{B}\right]_i$$

$$= \sum_{a,b} \frac{e_a{}^2 e_b}{m_a} \int \nu_{ab}{}^{(-)} \frac{r_i}{r^3} \, dr^3$$

If the ions and electrons are not too far out of equilibrium with each other, the ion and electron stress tensors are of the same order of magnitude. Then since $n_+$ and $n_-$ are, under the same condition, also of the same order of magnitude, the ion-pressure term is insignificant compared with the electron-pressure term, since $m_-/m_+ \ll 1$. Now from Equations 4.21 and 4.22,

$$\frac{\vec{u}_+}{m_+} + \frac{\vec{u}_-}{m_-} \approx \frac{1}{m_-}\left(\vec{u} - \frac{\vec{J}}{en_-}\right)$$

After one makes this substitution and drops the ion-pressure term, the generalized Ohm's law becomes

$$\frac{\partial J_i}{\partial t} + u_k \frac{\partial J_i}{\partial x_k} + J_k \frac{\partial u_i}{\partial x_k} + J_i \frac{\partial u_k}{\partial x_k} - \frac{e}{m_-} \frac{\partial P_{-ik}}{\partial x_k} - \frac{e^2 n_-}{m_-} \left(E_i + (\vec{u} \times \vec{B})_i\right)$$

$$+ \frac{e}{m_-} (\vec{J} \times \vec{B})_i = \sum_{a,b} \frac{e_a{}^2 e_b}{m_a} \int \nu_{ab}{}^{(-)} \frac{r_i}{r^3} \, dr^3 \qquad (4.28)$$

In Spitzer's development of Equation 4.28,[47] it is clear that the collision integral on the right side represents the net exchange of momentum between ions and electrons. To close the system of equations, one must approximate this integral by a phenomenological argument. When electrical neutrality is not assumed, Spitzer's form for the collision integral is

$$\sum_{a,b} \frac{e_a^{\,2} e_b}{m_a} \int \nu_{ab}^{(-)} \frac{\vec{r}}{r^3}\, dr^3 = -\frac{e^2 n_-}{m_-} \frac{\vec{J}}{\sigma} \tag{4.29}$$

in which $\sigma$ is called the electrical conductivity.

Green[46] states that this form is "not in agreement with a more fundamental analysis," and he briefly outlines that analysis. Unfortunately, he does not propose a "more correct" phenomenological equation for the collision integral; hence, in this sense he has not completely closed his set of magnetohydrodynamic equations. At least in the case of collision-dominated shock waves, however, the form of the right side of Equation 4.29 — with a scalar resistivity $1/\sigma$ — is a physically reasonable one and is accepted in the remainder of this monograph.

## 2. Specialization of the Magnetohydrodynamic Equations to One-Dimensional Steady Flow

For continuum magnetohydrodynamics, the complete equations consist of the continuity equation (4.1), three momentum equations (4.9), an energy equation (4.11), seven Maxwell equations (4.4, 4.5, 4.6), and the three components of the generalized Ohm's law (4.28). These are fifteen equations in the fifteen unknowns $\rho$, $T$, $\vec{u}$, $\vec{J}$, $\vec{E}$, $\vec{B}$, $\rho_c$ if the pressure tensor, heat-flux vector, and collision integral in Equation 4.28 are written in terms of the other unknowns. When the process is collision-dominated, the stress tensor can be approximated by the isotropic form (Equation 4.12), the heat-flux vector by the form of Equation 4.25 with scalar coefficients, and the collision integral by Equation 4.29 with a scalar coefficient of electrical conductivity.

For one-dimensional steady flow in which all dependent variables depend only on the x-coordinate, it is evident that Equations 4.1, 4.9, and 4.11 can be immediately integrated. They become

$$\rho u = G \tag{4.30}$$

which can also be written

$$u = G\tau \tag{4.30a}$$

if $\tau = 1/\rho$,

$$P_{xx} + \rho u^2 - \frac{1}{2}\epsilon E_x{}^2 + \frac{1}{2}\epsilon E_y{}^2 + \frac{1}{2}\epsilon E_z{}^2 - \frac{1}{2}\frac{B_x{}^2}{\mu} + \frac{1}{2}\frac{B_y{}^2}{\mu} + \frac{B_z{}^2}{2\mu} = F_x^* \tag{4.31}$$

$$P_{xy} + \rho uv - \epsilon E_x E_y - \frac{B_x B_y}{\mu} = F_y \tag{4.32}$$

$$P_{xz} + \rho uw - \epsilon E_x E_z - \frac{B_x B_z}{\mu} = F_z \tag{4.33}$$

$$\rho u \left(U + \frac{u^2 + v^2 + w^2}{2}\right) + P_{xx}u + P_{xy}v + P_{xz}w + Q_x + \frac{E_y B_z}{\mu} - \frac{E_z B_y}{\mu} = GH \tag{4.34}$$

in which G, $F_x^*$, $F_y$, $F_z$, H are constants and u, v, w represent $u_x$, $u_y$, $u_z$, respectively. Since Equation 4.7 shows that $B_x$ = const, the substitution

$$F_x = F_x^* + \frac{1}{2}\frac{B_x{}^2}{\mu}$$

will be made in the final equations.

The required components of the pressure tensor can be found from Equation 4.12. They are

$$P_{xx} = p \quad m_1 \frac{du}{dx} \tag{4.35}$$

$$P_{xy} = - m_2 \frac{dv}{dx} \tag{4.36}$$

$$P_{xz} = - m_2 \frac{dw}{dx} \tag{4.37}$$

in which

$$m_1 = \frac{4}{3}\eta + \zeta \qquad m_2 = \eta \tag{4.38}$$

in accordance with a notation adopted by Germain.[13]

Only the x-component of the heat flux appears in the above steady-state equations. From Equation 4.25 it is

$$Q_x = -\kappa \frac{dT}{dx} - \kappa' J_x \tag{4.39}$$

The internal energy U and scalar pressure p must be expressed in terms of the other state variables if the system of equations is to close. Assuming that the gas is ideal,

$$\left.\begin{aligned} U &= \frac{RT}{\gamma - 1} \\ p &= \frac{RT}{\tau} \end{aligned}\right\} \tag{4.40}$$

**From Ampère's law (Equation 4.4),**

$$j_x \equiv J_x + \rho_c u = 0 \tag{4.41}$$

$$j_y \equiv J_y + \rho_c v = -\frac{1}{\mu}\frac{dB_z}{dx} \tag{4.42}$$

$$j_z \equiv J_z + \rho_c w = \frac{1}{\mu}\frac{dB_y}{dx} \tag{4.43}$$

Faraday's law of induction (Equation 4.5) can be integrated directly, producing the result that

$$E_y = \text{const}, \quad E_z = \text{const} \tag{4.44}$$

The charge density is given by Equation 4.6 as

$$\rho_c = \epsilon \frac{dE_x}{dx} \tag{4.45}$$

and the auxiliary condition (Equation 4.7) gives

$$B_x = \text{const} \tag{4.46}$$

Finally, if it is assumed that electron viscosity is negligible, the generalized Ohm's law (Equation 4.28), with Equation 4.29, becomes

$$u\frac{dJ_x}{dx} + 2J_x\frac{du}{dx} - \frac{e}{m_-}\frac{dp_-}{dx} - \frac{e^2 n_-}{m_-}(E_x + vB_z - wB_y)$$

$$+ \frac{e}{m_-}(J_y B_z - J_z B_y) = -\frac{e^2 n_-}{m_-}\frac{J_x}{\sigma} \tag{4.47}$$

$$u\frac{dJ_y}{dx} + J_x\frac{dv}{dx} + J_y\frac{du}{dx} - \frac{e^2 n_-}{m_-}(E_y + wB_x - uB_z)$$

$$+\frac{e}{m_-}(J_z B_x - J_x B_z) = -\frac{e^2 n_-}{m_-}\frac{J_y}{\sigma} \qquad (4.48)$$

$$u\frac{dJ_z}{dx} + J_x\frac{dw}{dx} + J_z\frac{du}{dx} - \frac{e^2 n_-}{m_-}(E_z + uB_y \quad vB_x)$$

$$+\frac{e}{m_-}(J_x B_y - J_y B_x) = -\frac{e^2 n_-}{m_-}\frac{J_z}{\sigma} \qquad (4.49)$$

In these equations, $n_-$ and $p_-$ must be expressed in terms of the single-fluid dependent variables. From Equations 4.16 and 4.18, and taking into account that both $m_-/m_+$ and $(n_+ - n_-)/n_-$ are small,

$$n_- \approx \frac{\rho}{m_+} - \frac{\rho_c}{e} \approx \frac{\rho}{m_+} = \frac{1}{m_+\tau} \qquad (4.50)$$

If the electrons are not too far out of equilibrium with the ions, and Equation 4.40 is used,

$$p_- \approx \frac{1}{2}p = \frac{1}{2}\frac{RT}{\tau} \qquad (4.51)$$

Now consider Equations 4.44. If a transformation is made to a new reference frame traveling at a constant transverse velocity $\vec{V}_t$ (perpendicular to the x-axis), the transformation of the electrical field is[51]

$$\vec{E}_t{}' = \vec{E}_t + \vec{V}_t \times \vec{B}$$

or

$$E_y{}' = E_y + V_z B_x \,, \quad E_z{}' = E_z - V_y B_x$$

Thus, if $B_x$(= const) is not zero, it is always possible to choose $V_y$ and $V_z$ so that $E_y{}' = E_z{}' = 0$. Then, from Equations 4.48 and 4.49, it is seen that outside the shock, where the current is zero, a reference frame can be chosen in such a way that

$$\frac{B_x}{u} = \frac{B_y}{v} = \frac{B_z}{w}$$

i.e., the magnetic-field and velocity vectors are parallel. This transformation was used by de Hoffmann and Teller.[2]

Many terms drop out of the above equations if $E_y = E_z = 0$; hence, without loss of generality a reference system is chosen in which this is true. This reference frame is then rotated so that at $x = \infty$ the velocity and magnetic-field vectors lie in the x-y plane. Then, referring to Equation 4.33, $w = B_z = 0$ at $x = \infty$. But since $P_{xz} = E_x = 0$ outside the shock, $F_z = 0$; and since $\vec{v}$ and $\vec{B}$ are parallel outside the shock, it follows that $w = B_z = 0$ at both $x = \pm\infty$ but not necessarily through the shock. This is the proof that a shock considered as a discontinuity in a perfect fluid is two-dimensional.

The fifteen macroscopic equations now reduce to ten equations if, in addition to taking the above transformation into account, we consider Equation 4.46 and eliminate $u$ and $\rho_c$ by means of Equations 4.30a and 4.41. In the order 4.31, 4.32, 4.34, 4.43, 4.49, 4.45, 4.47, 4.33, 4.42, 4.48, the ten equations of the shock layer then become

$$m_1 G \frac{d\tau}{dx} = \frac{RT}{\tau} + G^2\tau + \frac{B_y^{\,2}}{2\mu} + \boxed{\frac{B_z^{\,2}}{2\mu}} - \frac{\epsilon E_x^{\,2}}{2} - F_x \qquad (4.52)$$

$$m_2 \frac{dv}{dx} = G v - \frac{B_x}{\mu} B_y - F_y \qquad (4.53)$$

$$\frac{\kappa}{G}\frac{dT}{dx} = \frac{\gamma RT}{\gamma - 1} + \frac{G^2\tau^2}{2} + \frac{1}{2}v^2 + \boxed{\frac{1}{2}w^2} - m_1 G\tau \frac{d\tau}{dx}$$

$$- \frac{m_2}{G} v \frac{dv}{dx} - \boxed{\frac{m_2}{G} w \frac{dw}{dx}} - \frac{\kappa'}{G} J_x - H \qquad (4.54)$$

$$\frac{1}{\mu}\frac{dB_y}{dx} = J_z - \boxed{\frac{J_x w}{G\tau}} \qquad (4.55)$$

$$G\tau \frac{dJ_z}{dx} + \boxed{J_x \frac{dw}{dx}} + G J_z \frac{d\tau}{dx} - \frac{e^2}{m_+ m_- \tau}(u B_y \quad v B_x)$$

$$+ \boxed{\frac{e}{m_-}(J_x B_y - J_y B_x)} = \frac{-e^2}{m_+ m_- \tau}\frac{J_z}{\sigma} \qquad (4.56)$$

. . . . . . . . . . . . . . . . . . . . . . . . . . . . . . .

$$\epsilon \frac{dE_x}{dx} = -\frac{J_x}{G\tau} \tag{4.57}$$

$$G\tau \frac{dJ_x}{dx} + 2GJ_x \frac{d\tau}{dx} - \frac{e}{2m_-}\frac{d}{dx}\left(\frac{RT}{\rho}\right) - \frac{e^2}{m_+m_-\tau}\left(E_x + \boxed{vB_z - wB_y}\right)$$

$$+ \frac{e}{m_-}\left(\boxed{J_yB_z} - J_zB_y\right) = -\frac{e^2}{m_+m_-\tau}\frac{J_x}{\sigma} \tag{4.58}$$

. . . . . . . . . . . . . . . . . . . . . . . . . . . . . . .

$$m_2 \frac{dw}{dx} = Gw - \frac{B_x}{\mu} B_z \tag{4.59}$$

$$\frac{1}{\mu}\frac{dB_z}{dx} = -J_y + \frac{J_x v}{G\tau} \tag{4.60}$$

$$G\tau \frac{dJ_y}{dx} + J_x \frac{dv}{dx} + GJ_y \frac{d\tau}{dx} - \frac{e^2}{m_+m_-\tau}(w B_x - G\tau B_z)$$

$$+ \frac{e}{m_-}\left(J_zB_x - \boxed{J_xB_z}\right) = -\frac{e^2}{m_+m_-\tau}\frac{J_y}{\sigma} \tag{4.61}$$

The reason for this particular ordering of the equations, the dotted boxes, and the separation into three groups will be explained in the next section.

## 3. Dimensional Analysis and Simplification of the Shock-Layer Equations

The primary motivation for this analysis has been to extend work of others to include the effect of current inertia; hence, the first topic in this section is a simple argument which shows that the inertia and collision terms in the current equations are of the same order of magnitude. We then show that the terms in dotted boxes in Equations 4.52-4.61 are negligible but that the electric force in Equation 4.52 is important.

Importance of Current Inertia. The electrical conductivity $\sigma$ can be related to the collision frequency $\nu_c$ by the simple formula

$$\nu_c = \frac{n_- e^2}{m_- \sigma} \tag{4.62}$$

The cyclotron frequency $\omega_c$ and the plasma frequency $\omega_p$ will also be needed later. They are

$$\omega_c = \frac{eB}{m_-} \tag{4.63}$$

$$\omega_p = \sqrt{\frac{n_- e^2}{m_- \epsilon}} \tag{4.64}$$

These formulas are given by Spitzer[47] but have been converted to MKS units, and, in Equation 4.62, the substitution $\eta = 1/\sigma$ has been made.

In Equations 4.47, 4.48, and 4.49, note that the right sides can be written $-\nu_c J_i$. Then, compare this collision term with the first term on the left side. For example, consider

$$\left(u \frac{d}{dx} + \nu_c\right) J_z$$

The current is zero outside the shock (see Equations 4.41-4.43); hence, as an idealization, let the current profile be as represented in Figure 4.1. Then

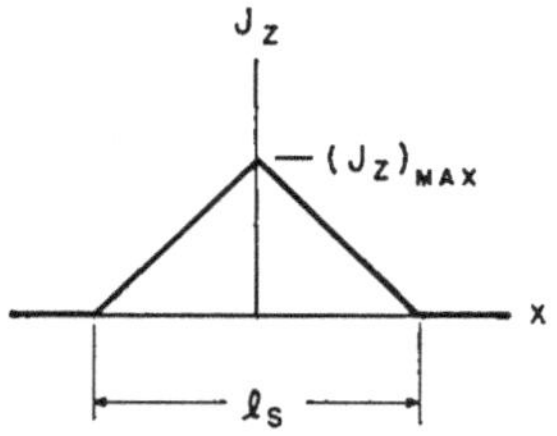


Figure 4.1. Idealization of the current profile

$$\frac{dJ_z}{dx} = 2\frac{(J_z)_{max}}{\ell_s}$$

where $\ell_s$ is the shock thickness. Hence,

$$u\frac{d}{dx} + \nu_c = \frac{2u}{\ell_s} + \nu_c$$

But the collision frequency is $\nu_c = u_m/\ell_m$, where $u_m$ is the mean molecular velocity and $\ell_m$ the mean free path. By simple kinetic theory, $u_m$ is roughly 20 per cent higher than the speed of sound, a, and except for some slow shocks, $u > a$ ahead of the shock. In collision-dominated shocks, $\ell_s$ is usually a few times greater than $\ell_m$; thus, $2u/\ell_s$ is of the same order of magnitude as $\nu_c$. In shocks reported to be much thinner than a mean free path,[41] the inertia terms dominate. Because the profile depicted in Figure 4.1 is least favorable to the inertia term, it is clearly never possible a priori to neglect current inertia.

Qualitative Discussion of the Shock-Layer Equations. Consider Equations 4.52 through 4.56. If we neglect the terms in dotted boxes, the electric term $(1/2)\epsilon E_x^2$ in Equation 4.52, the term proportional to $\kappa'$ in Equation 4.54, and the first and third terms in Equation 4.56 (due to current inertia), we obtain the system treated by Germain[13] and others. Note in particular that $J_z$ is the most important current component since it generates the change in $B_y$ across the shock. The first five dotted boxes contain terms dependent upon $w$ and $B_z$. We expect these terms to be small, because $w = B_z = 0$ outside the shock, and we shall show that they are of the order of $u^2/c^2$ compared with other terms. The second dotted box in Equation 4.56 is the z-component of the Hall current and depends on the secondary currents $J_x$ and $J_y$. These terms are proportional to the cyclotron frequency $\omega_c (= eB/m_-)$, and hence, in keeping with a basic assumption of this work, will be dropped. With all of the boxed terms removed, Equations 4.52-4.56 would be a closed system except for the electric-force term in Equation 4.52 and the heat-flux component $\kappa' J_x$ in Equation 4.54. It will be shown that the electric force is not negligible inside of the shock, a result which is not surprising, since charge separation is intuitively evident in shocks in ionized gases.

The second group of shock-layer equations, Equations 4.57 and 4.58, are linear differential equations for $E_x$ and $J_x$, forced by the electron-pressure gradient and the Hall current proportional to $J_z$. We shall show that the first boxed terms are of the order of $u^2/c^2$ compared with $E_x$. The second boxed term is a Hall current of the order of $\omega_c/\nu_c$ compared with the collision term on the right side, because $J_x$ and $J_y$ are of the same order of magnitude. The second part of the Hall current is not necessarily small, because it will turn out that $J_z >> J_x$. Note that the coefficients and forcing function in Equations 4.57 and 4.58 depend only on the variables of Equations 4.52-4.56. The two quantities $E_x$ and $J_x$ then feed back into the primary system through the quadratic term $(1/2)\epsilon E_x^2$ and through $\kappa' J_x$.

The third group, Equations 4.59-4.61, is a set of linear differential equations for the variables $w$, $B_z$, and $J_y$. These equations are forced by terms proportional to $J_x$ in Equations 4.60 and 4.61 and by the Hall-current term $(e/m_-)J_z B_x$ in Equation 4.61. Their feedback into Equations 4.52-4.58 is negligible; thus, they need not be considered in a study of the existence and primary qualitative behavior of shocks.

Quantitative Estimates of the Small Terms. The components $J_x$ and $J_y$ of the conduction current depend on the existence of the electron-pressure gradient in Equation 4.58 and on the Hall-current components proportional to $J_z$ in Equations 4.58 and 4.61. If these two forcing functions were zero, Equations 4.57-4.61 would be a homogeneous linear set of differential equations, the finite solutions of which must become unbounded at either plus or minus infinity. Since this is not physically acceptable, the homogeneous solution must vanish, and only particular solutions due to the forcing functions can exist; thus $J_x$ and $J_y$ arise from essentially the same causes, and their relative magnitudes must depend primarily on the obliqueness of the shock. Because it will turn out that the terms to be neglected will be of the order of $u^2/c^2$ compared with other terms, it is then reasonable to make the loose approximation that

$$= O(J_x) \tag{4.65}$$

in which O means "of the order of."

It is easiest first to consider the Hall currents. Since $J_z$ produces the change in $B_y$ across the shock, i.e., the principal magnetohydrodynamic effect, it must be true that $J_z > J_x$, $J_y$. This will be justified more fully later. Then, in Equation 4.56 the Hall current — in the second dotted box — is small compared with the collision term on the right side because of the assumption $\omega_c << \nu_c$. Similarly, in Equations 4.58 and 4.61, because of Equation 4.65, the boxed Hall current is small; but since at this point it is not known how much larger $J_z$ is than $J_x$ and $J_y$, the Hall current proportional to $J_z$ is kept.

Now, there must be a length L of the order of the shock thickness such that

$$\frac{dB_z}{dx} = O\left(\frac{B_z}{L}\right)$$

Let us consider Equation 4.60. First, $G\tau = u = O(v)$ for oblique shocks. This is a rough approximation, and the two could differ by almost two orders of magnitude; but even this is close enough for the present analysis. Then,

$$B_z = O(\mu L J_x) \tag{4.66}$$

and similarly from Equation 4.57,

$$E_x = O\left(\frac{LJ_x}{\epsilon u}\right) \tag{4.67}$$

Hence,

$$\epsilon E_x^2 = O\left(\frac{L^2 J_x^2}{\epsilon u^2}\right)$$

and

$$\frac{B_z^{\,2}}{\mu} = O(\mu L^2 J_x^{\,2}) = O\left(\frac{L^2 J_x^{\,2}}{\epsilon u^2} \frac{u^2}{c^2}\right) \ll \epsilon E_x^{\,2} \tag{4.68}$$

in which $c^2 = 1/\mu\epsilon$. Thus, since $u^2/c^2 \ll 1$ in a nonrelativistic shock, it is clear that the boxed term in Equation 4.52 is very much smaller than the electric-force term, and can be dropped.

In the next few paragraphs, we shall prove that $(1/2)\,\epsilon E_x^{\,2}$ can be of the order of the other terms in Equation 4.52; hence, Equation 4.68 shows that

$$B_z^{\,2} \ll B_y^{\,2} \tag{4.69}$$

Equation 4.59 shows that $w$ is forced by $B_z$. Both are zero at $\pm\infty$ and hence must be somewhat in the nature of pulse functions. The transverse velocity $w$ attains its peak when $dw/dx = 0$; so it is reasonable to estimate that

$$w = O\left(\frac{B_x}{\mu} \frac{B_z}{\rho u}\right)$$

and consequently that

$$w^2 = O\left(\frac{B_x^{\,2}}{\mu\rho u^2} \frac{B_z^{\,2}}{\mu\rho}\right) \tag{4.70}$$

But the first factor in Equation 4.70 is the square of the Alfvén velocity divided by the flow velocity. On the same level of approximation as used above,

$$\frac{B_x^{\,2}}{\mu\rho} = O(u^2) = O(v^2) = O\left(\frac{B_y^{\,2}}{\mu\rho}\right) \tag{4.71}$$

if magnetohydrodynamic effects are to be produced. Next,

$$w^2 = O\left(\frac{B_z^{\,2}}{\mu\rho}\right)$$

and, in combining this with Equations 4.69 and 4.71, it is clear that

$$w^2 \ll v^2 \tag{4.72}$$

This permits elimination of the boxed terms in Equation 4.54.

Then, in Equation 4.55,

$$\frac{J_x w}{u} = O\left(\frac{B_x B_z}{\mu \rho u^2} \frac{B_z}{\mu L}\right) << \frac{B_y}{\mu L} = O(J_z) \tag{4.73}$$

Using Equations 4.66, 4.69, and 4.73,

$$J_z >> J_x, J_y$$

Hence, the boxed term in Equation 4.55 and also the remaining one in Equation 4.56 are negligible. Similarly, the first boxed term in Equation 4.58 is negligible.

Quantitative Estimate of the Electric Force. Consider Equations 4.57 and 4.58. They may now be estimated by

$$\epsilon E_x = O\left(\frac{J_x L}{u}\right)$$

$$\frac{3uJ_x}{L} + \nu_c J_x - \epsilon \omega_p^2 E_x = O\left(\frac{en_- kT}{m_- L} + J_z \omega_c\right)$$

in which the definitions of three frequencies, $\nu_c$, $\omega_p$, and $\omega_c$ (Equations 4.62-4.64), have been used. Eliminating $J_x$, assuming that $u/L = O(\nu_c)$ from page 102, and using Equation 4.73, the following estimate results:

$$(4\nu_c^2 - \omega_p^2)\epsilon E_x = O\left(\frac{en_- kT}{m_- L} + \frac{m_-}{\mu L e}\omega_c^2\right)$$

$$= O\left[\frac{m_-}{\mu L e}\left(\omega_p^2 \frac{kT}{m_- c^2} + \omega_c^2\right)\right] \tag{4.74}$$

The quantity $kT/m_- c^2$ is the ratio of thermal energy to the rest energy of the electron and is very small in a nonrelativistic plasma. The plasma frequency can be larger or smaller than the collision frequency in the region of interest; hence, even though $\omega_p^2 >> \omega_c^2$, the small coefficient $kT/m_- c^2$ may cause the Hall-current term $(\omega_c^2)$ to dominate over the electron-pressure term. Thus, the assumption $\omega_c << \nu_c$ does not necessarily eliminate Hall effects.

To estimate $\epsilon E_x^2$ from Equation 4.74, four possible cases are considered. In the first one, assume that $\nu_c^2 >> \omega_p^2$ and $\omega_c^2 << \omega_p^2\ (kT/m_- c^2)$. Then, letting $p_- = n_- kT$ and using the first form of the right side of Equation 4.74, we can write

$$\epsilon E_x^2 = O\left(\frac{e^2 n_-}{\nu_c^2 m_- \epsilon}\ \frac{kT}{m_-(\nu_c L)^2}\ p_-\right)$$

Substituting from Equation 4.64 and letting $u = O(\nu_c L)$, we have

$$\epsilon E_x^2 = O\left[\left(\frac{\omega_p}{\nu_c}\right)^2 \left(\frac{kT}{m_+ u^2}\right)\left(\frac{m_+}{m_-}\right) p_-\right]$$

The first factor is small by assumption. The second is the ratio of thermal energy to the directed kinetic energy of the flow with respect to the shock, and is small ahead of the shock but large behind it. The third factor is of the order of 2000, and the fourth is of the order of the total pressure. Thus the factor in front of $p_-$ is certainly not a priori small.

In the second case, let us assume that $\nu_c^2 << \omega_p^2$ and that $\omega_c^2 << \omega_p^2\ (kT/m_- c^2)$. Then,

$$\epsilon E_x^2 = O\left[\left(\frac{\nu_c}{\omega_p}\right)^2 \left(\frac{kT}{m_+ u^2}\right)\left(\frac{m_+}{m_-}\right) p_-\right]$$

Thus, reversing the relative magnitudes of $\nu_c$ and $\omega_p$ leaves one with an identical estimate. Clearly, when $\nu_c = O(\omega_p)$, the mass ratio dominates, and the electrical pressure can be much greater than the gas pressure.

In the third example, let us assume that $\nu_c^2 >> \omega_p^2$ and that $\omega_c^2 >> \omega_p^2\ (kT/m_- c^2)$. Then after some manipulation,

$$\epsilon E_x^2 = O\left[\left(\frac{c}{u}\right)^2 \left(\frac{\omega_c}{\nu_c}\right)^2 \frac{B_y^2}{\mu}\right]$$

The first term is very large and probably predominates over the small second term. Therefore, in this case the electric pressure may be considerably greater than the magnetic pressure.

In the fourth and final case, assume that $\nu_c^2 << \omega_p^2$ and $\omega_c^2 >> \omega_p^2\ (kT/m_- c^2)$. Then,

$$\epsilon E_x^2 = O\left[\left(\frac{\nu_c}{\omega_p}\right)^4 \left(\frac{c}{u}\right)^2 \left(\frac{\omega_c}{\nu_c}\right)^2 \frac{B_y^2}{\mu}\right]$$

in which case $\epsilon E_x^2$ is smaller than in the third case by the factor $(\nu_c/\omega_p)^4$.

From the above analysis it is clear that the electric pressure can be an important contribution to Equation 4.52.

## 4. The Final Form of the Shock-Layer Equations

After eliminating the boxed terms in Equations 4.52-4.61, w, $B_z$, and $J_y$ no longer couple into the first seven equations. Hence, Equations 4.59-4.61, from which they are derived, can be dropped from consideration.

If the derivatives on the right side of Equation 4.54 are eliminated by means of Equations 4.52 and 4.53, the five primary shock-layer equations can be written

$$m_1 G \frac{d\tau}{dx} = \frac{RT}{\tau} + G^2\tau + \frac{B_y^2}{2\mu} \quad F_x - \frac{\epsilon E_x^2}{2} \tag{4.75}$$

$$\frac{m_2}{G}\frac{dv}{dx} = v - \frac{B_x}{\mu G} B_y - \frac{F_y}{G} \tag{4.76}$$

$$\frac{\kappa}{G}\frac{dT}{dx} = \frac{RT}{\gamma-1} - \frac{G^2\tau^2}{2} - \frac{1}{2}v^2 - \frac{\tau B_y^2}{2\mu} + \frac{B_x}{\mu G} vB_y$$

$$+ \tau F_x + \frac{F_y}{G} v + \tau \frac{\epsilon E_x^2}{2} - \frac{\kappa'}{G} J_x - H \tag{4.77}$$

$$\frac{1}{\mu}\frac{dB_y}{dx} = J_z \tag{4.78}$$

$$\nu G\tau^2 \frac{dJ_z}{dx} = -\nu \tau G J_z \frac{d\tau}{dx} + G\tau B_y - vB_x - \frac{J_z}{\sigma} \tag{4.79}$$

in which the number $\nu = m_+ m_-/e^2$. In this form it can easily be seen that if $\nu = \kappa' = \epsilon E_x^2 = 0$, the above equations are almost the same as those analyzed by Germain.[13] The only incidental difference is that he transformed to a reference frame in which $F_y = 0$ but $E_z \neq 0$ instead of vice versa.

Finally, after substituting $J_x$ from Equation 4.57 and $J_z$ from Equation 4.78 into Equation 4.58, the following second-order linear differential equation is obtained for $E_x$:

$$\frac{d^2}{dx^2}\epsilon E_x + \frac{1}{u}\left(\nu_c + 3\frac{du}{dx}\right)\frac{d\epsilon E_x}{dx} + \frac{\omega_p^2}{u^2}\epsilon E_x = -\frac{e}{m_- u^2}\frac{d}{dx}\left(p_- + \frac{B_y^2}{2\mu}\right) \tag{4.80}$$

in which the definitions given by Equations 4.62, 4.64, and $u = G\tau$ have been used. Then $J_x$ is obtained from

$$J_x = - G\tau \frac{d}{dx} \epsilon E_x \tag{4.57}$$

Equation 4.80 has been written in a form which maximizes its physical significance. It is clear that $\epsilon E_x$ obeys a second-order linear differential equation, the solution of which has a variable natural wave length $2\pi u/\omega_p$, and a variable damping coefficient $\zeta = (\nu_c + 3u')/2\omega_p$. It is forced by the gradient of the sum of the electron pressure plus the magnetic pressure. From the analysis on page 102, $u'$, which is predominantly negative, can be of the order of $\nu_c$; therefore, inside the shock, the damping coefficient can be zero or even negative. Outside the shock, collisions provide positive damping, and the forcing function vanishes.

## Chapter 5

## GENERAL QUALITATIVE STUDY OF THE SHOCK LAYER

### 1. Method of Analysis

The set of equations formulated in Chapter 4 can be written symbolically in the form

$$\frac{dy_i}{dx} = F_i(y_1, y_2, \cdots y_n) \qquad (i = 1, \cdots, n) \qquad (5.1)$$

Lefschetz[52] refers to this type of differential system as autonomous because the $F_i$ do not depend on the independent variable. He gives a number of general theorems on autonomous systems in n dimensions, then attacks the two-dimensional problem in more detail. McLachlan[53] also treats the two-dimensional autonomous system in a more elementary but satisfactory manner and provides many helpful examples.

The reader familiar with the mathematical theory of autonomous systems, given in the above references — and also in many other places — will have no difficulty with treatments of shock structure of the type given for ordinary gas-dynamic shocks by Weyl,[15] von Mises,[54] and Gilbarg,[10] and for magnetohydrodynamic shocks by Marshall,[43] Ludford,[44] Germain,[13] and Kulikovskii and Liubimov.[45] Among these studies, Marshall's is three-dimensional, Germain's is four-dimensional, and all the others are two-dimensional. The present problem, as formulated in Chapter 4, is seven-dimensional, but it will be shown that for existence studies it can be cut to five. Hence, although the procedures used in analysis of two-dimensional systems must first be thoroughly understood, it is natural to study Germain's paper as a more direct basis for the present work. In fact, it is difficult to see how our analysis could have been made without the use of several general techniques given by Germain or without the inspiration received from reading his paper.

A brief description of the method of analysis follows. The first step is the location of the singular points, at which all of the $F_i$ in Equations 5.1 are zero. In the present problem there are at most four of them — already discussed in Chapter 2 and shown in Figures 2.5 and 2.6. Since we refer to these singular points frequently in this chapter and the next, we shall use the notation $SP_i$ for the $i^{th}$ singular point. The importance of these singular points is that they are the only points in the n-dimensional space of the $y_i$ through which more than one integral curve of Equations 5.1 can pass; i.e., they are the origins and terminal points for the integral curves. At all ordinary points, the solution of Equations 5.1 exists and is unique[55] for a class of functions $F_i$ much more general than required in the present problem.

The singular points are the intersections of the n "null surfaces" $F_i = 0$. Since the $F_i$ are known functions independent of x, the null surfaces are known stationary surfaces in the n-dimensional space of the $y_i$, and a knowledge of the shapes of these surfaces and the regions of n-space which they bound is very important for determining the behavior of the integral curves.

After locating the null surfaces and singular points, the next step is to linearize Equations 5.1 about each singular point. Thus, if the $F_i$ are expanded into Taylor series about a singular point, Equations 5.1 become

$$\frac{d}{dx}(y_i - y_{i_0}) = \sum_{k=1}^{n} \frac{\partial F_i}{\partial y_k}(y_k - y_{k_0}) + \cdot \qquad (i = 1, \cdot \cdot \cdot, n)$$

or

$$\frac{d\delta y_i}{dx} = A_{ik}\, \delta y_k \tag{5.2}$$

in which summation over the repeated index k is implied. From the theory of linear differential equations, it is always possible to find a matrix $M_{ik}$ which will simultaneously diagonalize both sides of Equation 5.2.

If no two eigenvalues are equal, the system then takes the form

$$\frac{d\delta y_i'}{dx} = \lambda_i\, \delta y_i' \qquad (i = 1, \cdot \cdot \cdot, n) \tag{5.3}$$

in which $\lambda_i$ is the $i^{th}$ eigenvalue and $\delta y_i'$ is the $i^{th}$ normal coordinate or eigenvector.

The solution of Equations 5.3 is

$$\delta y_i' = c_i e^{\lambda_i x} \tag{5.4}$$

in which $c_i$ is an arbitrary constant. The $j^{th}$ component of the $i^{th}$ eigenvector, in the original coordinate system, is then

$$\delta y_j = M_{j(i)}^{-1} c_{(i)} \tag{5.5}$$

in which the parentheses imply that there is no summation. If $\lambda_i$ is real and positive, $\delta y_i' \to 0$ at the singular point — as it must — only if $x \to -\infty$ there. But the velocity of the flow through the shock is assumed positive in the direction of positive x; hence, for this eigenvalue (and eigenvector) to correspond to a shock, this particular singular point is upstream of the shock. Similarly, if $\lambda_i$ is real and negative, the particular singular point in question is a downstream point. We must emphasize that a particular singular point could be an upstream point for one eigenvalue and a downstream point for another; in other words, points for which x increases can leave the singular point in the direction of one eigenvector but approach the singular point in the direction of another eigenvector. Clearly, then, determination of the signs of the eigenvalues will be an important consideration.

To see the general behavior of integral curves near a singular point, it is useful to study a two-dimensional system. Dividing the first of Equations 5.3 by the second, we obtain

$$\frac{d\eta_1}{d\eta_2} = \frac{\lambda_1}{\lambda_2}\frac{\eta_1}{\eta_2} = \mu\frac{\eta_1}{\eta_2} \tag{5.6}$$

in which $\eta_1 = \delta y_1'$, $\eta_2 = \delta y_2'$. Thus the independent variable x is eliminated, and Equation 5.6 represents a family of stationary curves in the $\eta_1$-$\eta_2$ plane. The solution of Equation 5.6 is

$$\eta_1 = C\eta_2^{\mu} \tag{5.7}$$

from which

$$\frac{d\eta_1}{d\eta_2} = C\mu\eta_2^{\mu-1}$$

Thus, if $\mu > 1$, $d\eta_1/d\eta_2 \to 0$ as $\eta_2 \to 0$ for all finite C. The exception to this is the case of a point approaching the singular point along the $\eta_1$-axis, i.e., in the direction of the first eigenvector. Then, for that point, $d\eta_1/d\eta_2 = \infty$ from Equation 5.6, and it therefore stays on the $\eta_1$-axis. Under these conditions, the singular point is called a "node" and is depicted in Figure

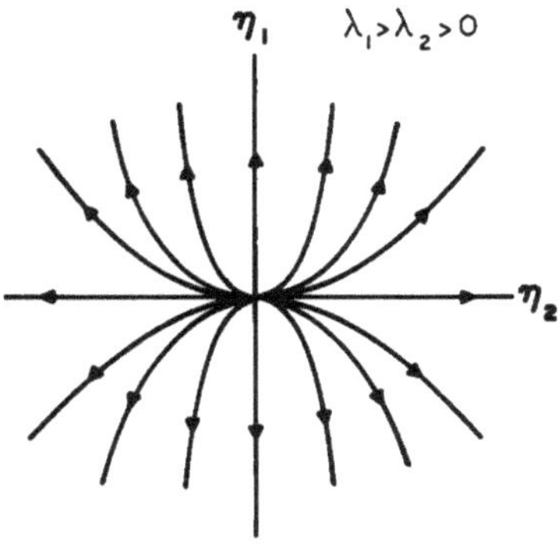


Figure 5.1. An upstream node

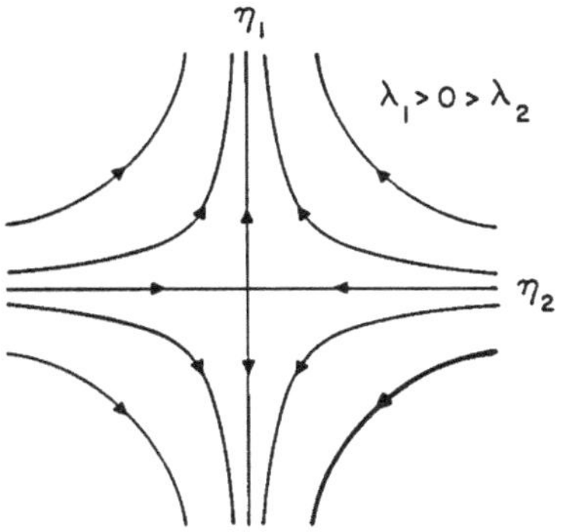


Figure 5.2. A saddle point

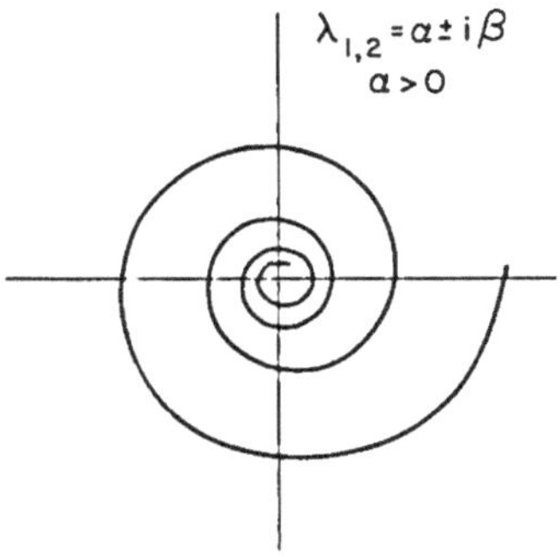


Figure 5.3. An upstream spiral point

5.1. The arrows point in the direction of increasing x. Note also that a single curve leaves in the direction of the eigenvector corresponding to the larger eigenvalue, whereas an infinity of integral curves leave in the direction corresponding to the smaller eigenvalue.

In the case $\lambda_1 > 0 > \lambda_2$ ($\mu < 0$), the solution, Equation 5.7, takes the form shown in Figure 5.2. It is called a "saddle point."

When $\lambda_1$, $\lambda_2$ are a pair of complex conjugates, it can be shown[53] that the integral curves take the form of a logarithmic spiral as depicted in Figure 5.3. It is an upstream spiral if $Re(\lambda_1, \lambda_2) > 0$ and a downstream spiral in the opposite case. In this instance there are no real eigenvectors; however, the plane in which the spiral lies is defined. For example, in the case of a five-dimensional system in which there are three real eigenvalues and a pair of complex-conjugate eigenvalues, the plane of the spiral must be orthogonal to the three-space defined by the three real orthogonal eivenvectors.

In a n-dimensional problem, it is clear from the form of Equation 5. 3 that each normal two-space can be viewed separately, and that one singular point can be a node in one plane, a saddle point in another, and a spiral in another, upstream and downstream eigenvectors being mixed as the case may be. From the character of the integral curves in each two-space it is possible to visualize how they look in any three-space, but of course it is not possible to visualize directly spaces of higher dimension.

The above discussion shows that is is possible to learn a great deal about the nature of the integral curves from the eigenvalues alone. One more property may also be estimated from them — the shock thickness. Once it has been determined which pair of eigenvalues $\lambda_A$, $\lambda_B$ corresponds to the actual shock-layer curve, it is clear from Equation 5.4 that at least an order-of-magnitude estimate of shock thickness is

$$\text{Shock thickness} \cong \frac{1}{\lambda_A} + \frac{1}{(-\lambda_B)} \tag{5.8}$$

in which $\lambda_A$ and $\lambda_B$ are the upstream and downstream eigenvalues, respectively.

After finding the signs of the eigenvalues (or of the real parts), the next step is to try to locate the eigenvalues with respect to the null surfaces, i.e., to find in which of the regions formed by the null surfaces the eigenvectors lie. This information can be found in general for the smallest and largest eigenvalues and sometimes can be inferred from orthogonality properties for intermediate eigenvalues; however, in the latter case the location of the eigenvectors usually depends on the relative values of the dissipation coefficients.

The next step is to analyze the integral curves as they cross the null surfaces to determine which way points moving in the direction of increasing x are going. In the formulation used by Germain[13] enough information was found up to this point to prove the existence and uniqueness of the fast shock, but not enough to obtain corresponding proofs for the intermediate and slow shocks.

A final and very important piece of information is the sign of the slope of the integral curves projected into a given plane. The sign is easy to determine, and since it is invariant in each of the known regions formed by the null surfaces, it is one of the most useful clues to the behavior of the integral curves.

Up to this point the study has been conducted in a five-dimensional space. Even with all the information obtainable, both near the singular points and in the large, it does not seem possible in this context to prove rigorously the existence and uniqueness properties of all shocks. There are several subcases of lower order which retain the principal difficulties inherent in the fifth-order system but in a simpler context; hence, these cases will be treated in the next chapter in order to increase understanding of the over-all problem. From all of this information, a very complete qualitative description of the properties of magnetohydrodynamic shocks can be obtained.

The next problem is to analyze the seventh-order system formulated in Chapter 4, and to show thereby that the coupling between the primary fifth-order system and the second-order system for the normal components of electric-field and conduction current need not be included in a study of shock existence.

## 2. Reduction to Fifth-Order System for Existence Studies

Consider the seven differential equations 4.75-4.80 and 4.57. First, neglect the coupling between the first five and the last two of these equations; i.e., let $\kappa' = 0$ and drop the term $\epsilon E_x^2$ in Equations 4.75 and 4.77. Then linearize the last two equations about the singular point at $E_x = J_x = 0$. The eigenvalues $\lambda$ of this linearized system are found directly from the homogeneous part of Equation 4.80. Thus

$$\lambda^2 + \frac{\nu_c}{u}\lambda + \left(\frac{\omega_p}{u}\right)^2 = 0 \tag{5.9}$$

The flow velocity $u > 0$ by assumption, and also $\nu_c > 0$; therefore, if the $\lambda$ are real, they are both negative, and if they are complex conjugates, their real part is negative. Hence, the singular point is either a downstream node or a downstream spiral point. A point moving along an integral curve in the negative x-direction starts at the origin and moves off to infinity; in particular, it cannot return to the origin.

For an integral curve to correspond to a shock-layer curve, its projection in the $J_x$-$E_x$ plane must begin at the origin on an outgoing eigenvector, then turn around, and come back to the origin. Hence, integral curves associated with the above two eigenvalues do not have the required character for a shock-layer curve.

Now consider the seventh-order system $(S_7)$ as a whole. When linearized about a singular point, it takes the form

$$\begin{bmatrix} M_5 & \kappa' C_{52} \\ \frac{e}{m_-} C_{25} & M_2 \end{bmatrix} \begin{vmatrix} y_5 \\ y_2 \end{vmatrix} = 0 \tag{5.10}$$

where $M_5$ is a $5 \times 5$ matrix representing the uncoupled fifth-order system $(S_5)$, $M_2$ is a $2 \times 2$ matrix representing the uncoupled second-order system $(S_2)$, $C_{25}$ and $C_{52}$ are the two coupling matrices, $y_5$ is the eigenvector of $S_5$, and $y_2$ is the eigenvector of $S_2$. When $\kappa' = 0$, Equations 5.10 can be written

$$M_5 \, y_5 = 0 \tag{5.11}$$

$$\frac{e}{m_-} C_{25} \, y_5 + M_2 \, y_2 = 0 \tag{5.12}$$

Thus, det $M_5 = 0$ gives the five eigenvalues of $S_5$, and Equation 5.12 shows that the seven-dimensional eigenvectors corresponding to these eigenvalues have components along $J_x$ and $E_x$. If det $M_2 = 0$ and det $M_5 \neq 0$, then $y_5 = 0$, and one arrives at the eigenvalues of Equation 5.9. Note that the eigenvectors corresponding to these latter two eigenvalues have components only along $J_x$ and $E_x$.

The above-described properties show that if there are any shock-layer curves in the case $\kappa' = 0$, they must leave and approach the singular points in the direction of one of the eigenvectors of $S_5$. Now allow $\kappa'$ to increase gradually. Then, the full determinant of Equations 5.10 must be solved in order to find the seven eigenvalues; however, it is clear, because of continuity, that the seven eigenvalues will move smoothly away from their values for $\kappa' = 0$. Similarly, the eigenvectors and in fact the whole field of integral curves will move gradually away from their positions for $\kappa' = 0$. The significance of this is that, at least for $\kappa'$ less than some maximum value $\tilde{\kappa}'$, any shock-layer curves present for $\kappa' = 0$ remain intact, although they will distort.

We shall assume in the following that $\kappa' < \tilde{\kappa}'$, i.e., that the coupling is small enough so that shock-layer curves existing for $\kappa' = 0$ are not broken. Then, for the following existence studies we can set $\kappa'$ equal to zero. The assumption that $\kappa'$ is small can be quite well justified by comparing the current coupling in Equation 4.77 with the internal-energy term. Thus, consider

$$\frac{RT}{\gamma - 1} - \frac{\kappa'}{G} J_x$$

The coefficient $\kappa'$ is given by Equation 4.26, and, from Reference 49, we can infer that the first term in that equation is the dominant one. Then, since $RT = nkT/\rho$, $\gamma = 5/3$ for a monatomic gas, and $G = \rho u$,

$$\frac{RT}{\gamma - 1} - \frac{\kappa' J_x}{G} \approx \frac{3}{2} RT \left(1 - \frac{5}{3} \frac{J_x}{enu}\right)$$

After substituting for $J_x$ from Equations 4.41 and 4.18, this expression becomes

$$\frac{3}{2} RT \left(1 + \frac{5}{3} \frac{n_+ - n_-}{n_+ + n_-}\right)$$

since $n = n_+ + n_-$. Thus, the current coupling term is of the order of the ratio of the charge separation to the total charge. This must be small because of the enormous electrical forces generated if it is not; hence, the effect of current coupling borders on being negligible entirely, and the weaker assumption that it is small is very reasonable.

Now, gradually replace the electric force in Equations 4.75 and 4.77 by allowing $\epsilon$ to increase from zero to its correct free-space value. First, notice that this term has no effect on the direction of approach of the integral curves to (or from) the singular points because it does not appear in the equations linearized about the singular points. Then, by continuity, if a particular shock-layer curve exists for $\epsilon = 0$, it must exist for $\epsilon = \delta\epsilon$, where $\delta\epsilon$ is an infinitesimal quantity; hence, assume that the shock-layer curve exists for $\epsilon = \epsilon_0$. If we can show that it also exists for $\epsilon_0 + \delta\epsilon$, we can prove by induction that it exists for all $\epsilon$.

In this discussion, remember that the pattern of integral curves around each singular point is independent of variations in $\epsilon$. Thus, if there were some $\epsilon$ for which the shock-layer curves cease to exist, they must go off to infinity for that $\epsilon$, because they cannot terminate at ordinary points and they cannot return to the singular points. If this is possible, it must be true that for some $\epsilon$ an infinitesimal change to $\epsilon + \delta\epsilon$ will cause a finite change in the position of the integral curve. But before the position of the integral curve can undergo a finite change, its slope must undergo a finite change. Equations 4.75 and 4.77 show that at every fixed point in $S_7$ an infinitesimal change in $\epsilon$ changes $d\tau$ and $dT$ by infinitesimal amounts and does not affect the other five components at all. This is true for all $\epsilon$; hence, all shock-layer curves which exist for $\epsilon = 0$ exist for all finite values of $\epsilon$. Alternately, consider the changes in the dependent variables of the shock-layer curve as $\epsilon$ changes to $\epsilon + \delta\epsilon$. This can be done by first assuming that all dependent variables in the shock-layer equations are functions of $\epsilon$, then expanding these equations in terms of the parameter $\delta\epsilon$. A set of linear differential equations with forcing functions proportional to $\delta\epsilon$ is obtained for the changes in the dependent variables. These equations have variable coefficients but can in principle be integrated from one singular point to another, i.e., from $x = -\infty$ to $x = +\infty$. The coefficients are integral functions of both $x$ and $\epsilon$; therefore, by the Cauchy-Kowalewski theorem,[55] the solutions are integral functions of $\epsilon$. Hence, the changes in the dependent variables, proportional to $\delta\epsilon$, are bounded for all $\epsilon$. QED

For the study of existence of shocks it is now clear that only Equations 4.75 to 4.79 need be considered. Dropping the $E_x$ and $J_x$ coupling terms, we can write these equations as follows in the standard form for autonomous systems:

$$m_1 G \frac{d\tau}{dx} = \frac{RT}{\tau} + G^2 \tau + \frac{B_y^2}{2\mu} - F_x \tag{5.13}$$

$$m_2 \frac{dv}{dx} = Gv - \frac{B_x}{\mu} B_y - F_y \tag{5.14}$$

$$\frac{\kappa}{G}\frac{dT}{dx} = \frac{RT}{\gamma - 1} - \frac{G^2\tau^2}{2} + F_x\tau - H - \frac{1}{2}v^2 - \frac{\tau B_y^2}{2\mu} + \frac{B_x}{\mu G} v B_y + \frac{F_y}{G} v \tag{5.15}$$

$$\frac{1}{\mu}\frac{dB_y}{dx} = J_z \tag{5.16}$$

$$\nu G \tau^2 \frac{dJ_z}{dx} = -\frac{\nu\tau}{m_1} J_z \left(\frac{RT}{\tau} + G^2\tau + \frac{B_y^2}{2\mu} - F_x\right) - \frac{J_z}{\sigma} + G\tau B_y - v B_x \tag{5.17}$$

## 3. Null Surfaces

The five null surfaces can be obtained by setting the right sides of each of Equations 5.13-5.17 equal to zero separately. Just as an ordinary surface is a curved two-space embedded in a Euclidean three-space, so these null surfaces are curved four-spaces embedded in a Euclidean five-space. The forms of the first four of these surfaces are known because they are given explicitly in terms of the five coordinates $\tau$, v, T, $B_y$, $J_z$; however, the fifth surface is not explicitly known because $m_1$ and $\sigma$ are unspecified functions of at least T. Since it is desirable to obtain existence proofs independent of the form of the dissipation coefficients, these functions will be left unspecified.

It is of course impossible to visualize surfaces directly in five-space, but visualization of these surfaces will be of great help in locating the shock-layer curves. The simple form of the null surface of Equation 5.16 is the first clue as to how a partial visualization can be achieved. It says that $J_z = 0$ at all singular points, and divides five-space into two semi-infinite regions — one for $J_z > 0$ and one for $J_z < 0$. Thus, as a first step, consider the projections of the null surfaces onto the Euclidean four-space $J_z = 0$. This affects only the form of the fifth null surface, which becomes

$$G\tau B_y - vB_x = 0 \tag{5.18}$$

The other three null surfaces were all cylindrical in the $J_z$-direction; hence, just as a picture of a circle can be imagined as the cross section of a cylinder, so a conception of these three surfaces can be obtained. The form of the fifth surface becomes much more complicated as $J_z$ moves away from zero; however, the fact that $J_z = 0$ at all singular points, and that the section of the fifth null surface through $J_z = \text{const}$ must depart from the form of Equation 5.18 in a continuous manner as $J_z$ increases, makes the projection onto the subspace $J_z = 0$ a useful tool, if it is not misunderstood.

The null surfaces projected onto $J_z = 0$ are now curved three-spaces embedded in a Euclidean four-space, and still cannot be visualized. The null surface

$$S(v, B_y) \equiv \mathbf{G}\mathbf{v} - \frac{\mathbf{B_x}}{\mu} \mathbf{B_y} \quad F_y = 0 \tag{5.19}$$

from Equation 5.14, however, suggests a further simplification. It is a plane in either of the three-spaces $v$-$B_y$-$\tau$ or $v$-$B_y$-$T$; hence, it is a Euclidean three-space which separates the four-space $v$-$B_y$-$\tau$-$T$ into two semi-infinte halves, one for $S > 0$ and one for $S < 0$. This is not difficult to visualize; thus, the projection of the surfaces from Equations 5.13, 5.15, and 5.18 onto the three-space $S(v, B_y) = 0$ is considered. This is tantamount to substitution of $v$ from Equations 5.19 into the remaining three surfaces, and it produces the following ordinary surfaces in the three-space $\tau$-$T$-$B_y$:

$$RT + G^2\tau^2 - F_x\tau + \frac{B_y^{\ 2}}{2\mu}\tau = 0 \tag{5.20}$$

$$\frac{RT}{\gamma - 1} - \frac{G^2\tau^2}{2} + F_x\tau - H + \frac{1}{2G^2}\left(\frac{B_x}{\mu} B_y + F_y\right)^2 - \tau\frac{B_y^{\ 2}}{2\mu} = 0 \tag{5.21}$$

$$B_y(\tau - \tau^*) = B_x F_y / G^2 \tag{5.22}$$

in which

$$\tau^* = B_x^{\ 2}/\mu G^2 \tag{5.23}$$

The first and last of these surfaces are identically those shown in Figures 2.1 and 2.2, respectively; but the second surface is not the surface shown in Figure 2.3. (The equation of the latter surface is obtained by eliminating $F_x$ and $F_y$ from Equation 5.21 by means of Equations 5.20 and 5.22.)

The energy surface, Equation 5.21, which is required for existence studies, may be visualized by rewriting it in the form

$$\frac{RT}{\gamma - 1} = \frac{G^2}{2}\tau^2 - F_x\tau + H - \frac{1}{2}\left(\frac{F_y}{G}\right)^2$$

$$+ (\tau - \tau^*)\frac{B_y^{\,2}}{2\mu}$$

$$- \frac{\tau^* F_y}{B_x} B_y \qquad (5.21a)$$

The first line represents a parabolic cylinder with generators parallel to the $B_y$-axis, the second line represents a warped parabolic surface with its curvature variable in the $\tau$-direction, and the third line represents a plane parallel to the $\tau$-axis. Each of these three surfaces is shown separately in Figure 5.4, from which the composite null surface is obtained by addition.

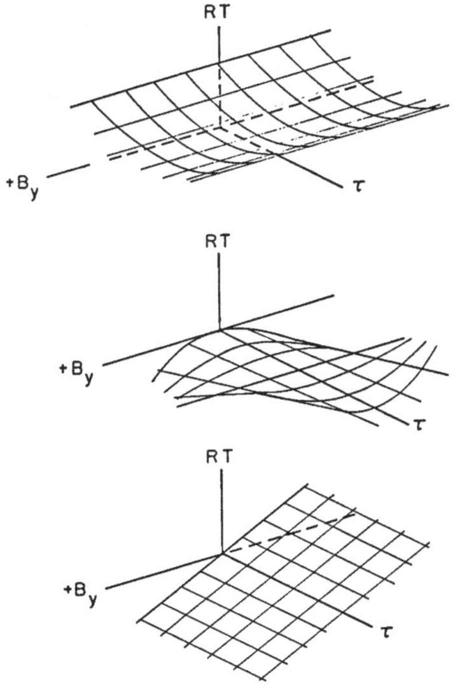

Figure 5.4. Components of the energy null surface

Consider the intersection of the energy surface with the normal-momentum surface. Since the middle portion of the energy surface, as shown in Figure 5.4, has downward curvature in some of the planes parallel to the $B_y$-$RT$ plane, the question is raised whether it is possible for the energy and normal-momentum surfaces to intersect without making contact in the plane $B_y = 0$. The following analysis will show that this can never happen.

We consider intersections of the surfaces of Equations 5.20 and 5.21 in the plane $B_y = 0$. Set $B_y$ equal to zero in these two equations, and eliminate RT between them. The result is the following quadratic equation in $\tau$:

$$\frac{1}{2}\left(\frac{\gamma+1}{\gamma-1}\right)G^2\tau^2 - \frac{\gamma F_x}{\gamma-1}\tau + H - \frac{F_y^{\,2}}{2G^2} = 0$$

This equation will always have real roots corresponding to real intersections if its discriminant D is always positive, where

$$D \equiv \left(\frac{\gamma}{\gamma-1}\right)^2 F_x^{\,2} - \left(\frac{\gamma+1}{\gamma-1}\right)(2G^2H - F_y^{\,2})$$

The constants G, $F_x$, $F_y$, H are given respectively by Equations 4.30, 4.31, 4.32, 4.34, and can be evaluated outside the shock layer. If a coordinate system is chosen so that $E_y = E_z = 0$, then $vB_x = uB_y$, and these constants become

$$G = \rho u$$

$$F_x = p + \rho u^2 + \frac{B_y^{\,2}}{2\mu}$$

$$F_y = \frac{B_y}{B_x}\left(\rho u^2 - \frac{B_x^{\,2}}{\mu}\right)$$

$$H = \frac{1}{\gamma-1}\frac{p}{\rho} + \frac{u^2}{2}\left(1 + \frac{B_y^{\,2}}{B_x^{\,2}}\right)$$

In the last equation, the internal-energy term has been placed in a form valid for perfect gases.

After substitution of these constants into D,

$$(\gamma-1)^2 D = \gamma^2\left(p + \rho u^2 + \frac{B_y^{\,2}}{2\mu}\right)^2 - (\gamma^2-1)\left(\frac{2\rho u^2 p}{\gamma-1} + \rho^2 u^4 + 2\rho u^2\frac{B_y^{\,2}}{\mu} - \frac{B_x^{\,2}B_y^{\,2}}{\mu^2}\right)$$

in which one cancellation in the second bracket has been made. This expression can be rearranged into the form

$$(\gamma-1)^2 D = \left(\gamma p - \rho u^2 + \frac{\gamma B_y^{\,2}}{2\mu}\right)^2 + (\gamma^2-1)\left(\frac{B_x^{\,2}B_y^{\,2}}{\mu^2} + 2\rho u^2 p\right)^2$$

$$+ (2-\gamma)(\gamma+1)\,\rho u^2\,\frac{B_y^{\,2}}{\mu}$$

For molecules with more than one degree of freedom, $1 < \gamma < 2$; hence, D is positive definite. This shows that the surfaces of Equations 5.20 and 5.21 always have real intersection points ($\tau_+ > \tau_- > 0$) in the plane $B_y = 0$. In terms of the constants $\tau_+$, $\tau_-$, the curve C of intersection of these surfaces is

$$\left(\frac{\tau + 1}{\gamma}\right) G^2 (\tau - \tau_+)(\tau - \tau_-) + \frac{\gamma B_y^{\ 2}}{2\mu} \left(\tau - \frac{\gamma - 1}{\gamma} \tau^*\right)$$

$$- (\gamma - 1) \frac{F_y \tau^*}{B_x} B_y = 0$$

For the three possible cases, it is shown in Figure 5.5.

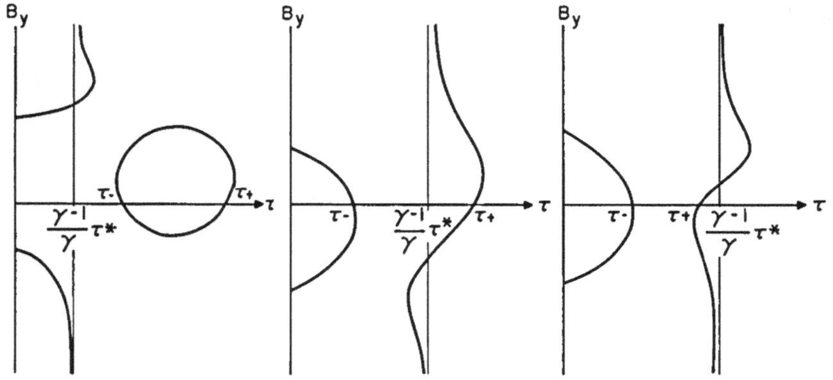


Figure 5.5. Intersection of the energy and normal-momentum null surfaces

Another aid in visualization of C and the regions around it can be obtained from the partial derivative of Equation 5.21 with respect to $\tau$. It is easy to see that the zeros of this derivative, $(\partial T/\partial \tau)B_y$, lie on the curve in $B_y$-$\tau$ space which results from Equation 5.20 when $T = 0$ and one factor $\tau$ is canceled. Thus, everywhere inside the base of the normal-momentum surface (see Figure 2.1), the slope $\partial T/\partial \tau$ of the energy surface is negative. Let C' be the portion of C on which $T > 0$, i.e., the portion of interest. Since the projection of C' on the $B_y$-$\tau$ plane lies within the base of the normal-momentum surface, it is clear that on C', negative variations $d\tau$ are always accompanied by positive variations dT.

We can gain further insight into the shape of the curve C from its slope $\partial B_y/\partial \tau$. If we eliminate T between Equations 5.20 and 5.21 and differentiate the resulting equation (the equation for the projection of C in the $B_y$-$\tau$ plane) with respect to $\tau$, we can express it in the form

$$\frac{\tau}{\mu}\frac{dB_y}{d\tau} = \frac{-(u^2 - a^2)}{B_y \tau + (\gamma - 1)\left[B_y(\tau - \tau^*) - \frac{F_y B_x}{G^2}\right]}$$

in which, from Equation 5.22, the term in parentheses vanishes on the projection of the transverse-momentum null surface in $T$-$B_y$-$\tau$ space. Hence, it is clear that at the singular points, the sign of $dB_y/d\tau$ is opposite to the sign of the factor $(u^2 - a^2)/B_y$. Making use of Inequalities 2.51, we can see that at the singular points upstream of a fast shock and downstream of a slow shock, this factor is positive, since $F_y$ in Equation 5.22 has been chosen positive. At the two middle singular points, the sign of the factor $u^2 - a^2$ is indefinite only for case 1, Figure 5.4. In cases 2 and 3, the point $u = a$ must occur in the imaginary region along the $\tau$-axis; hence, $dB_y/d\tau$ at the singular points is determined. Moving in the direction of positive $B_y$ from the point $B_y = 0$, $\tau = \tau_+$, it is now apparent that the first singular point (ahead of a fast shock) lies beyond the point $d\tau = 0$. Then, since $B_y(\tau - \tau^*) - F_y B_x/G^2 > 0$ on $C$ between the two fast-shock singular points, it is evident that as $\tau$ decreases between them, $T$ increases monotonically, and $B_y$ has a maximum point if and only if $u = a$ lies between the two points.

Consideration of the projection of the full null surfaces onto the plane $J_z = 0$ and then onto the plane $S(v, B_y) = 0$ (Equation 5.19) has merit as an aid in the analysis of behavior of integral curves only because all of the singular points lie in the spaces defined by these two equations. It is strictly correct, of course, only in the case $\nu = m_2 = 0$; nonetheless, it does prove to be useful.

We must emphasize again that the importance of the null surfaces is twofold. First, their simultaneous intersections define the four singular points which may be end points for shocks; and, second, they divide the five-dimensional configuration space into regions in which the slopes of the integral curves have certain definite and easily determinable directions. Moreover, the direction of the integral curves as $x$ increases is frequently easy to find on the null surfaces. In the next section we shall analyze the space immediately adjacent to the singular points. Then, in the following section we consider in detail the regions defined by the null surfaces, and develop methods for finding properties of the integral curves relative to them. Only after that can we consider the existence and uniqueness proofs.

## 4. The Linearized System and Its General Properties

It will be convenient to define new dimensionless variables, which vanish at the singular points, by means of the equations

$$\left.\begin{aligned}
\tau &= \tau_i\,(1 + \tau^*) \\
v &= v_i + u_i\, v^* \\
T &= T_i\,(1 + T^*) \\
B_y &= B_{y_i} + B_{x_i}\, B_y{}^* \\
J_z &= \frac{e u_i}{m_+ \tau_i}\, J_z{}^*
\end{aligned}\right\} \tag{5.24}$$

The subscript i indicates that the corresponding quantity is taken at the $i^{th}$ singular point. To obtain the linearized equations in the form desired, substitute Equations 5.24 into Equations 5.13-5.17, neglecting second and higher powers of the dimensionless variables and using Equations 5.18, 5.19, 5.20, and 5.22, which are valid at every singular point, to simplify the result. Then multiply Equations 5.13 and 5.14 by $\tau_i$, Equation 5.16 by $(m_- /e)G\tau_i{}^2$, and Equation 5.17 by $B_{y_i}/\mu G$. Next, eliminate $J_z{}^*$ from Equation 5.17 by substituting from Equation 5.16. Following this, use the substitutions $G = u_i/\tau_i$, $b_x{}^2 = \tau B_x{}^2/\mu$, $b_y{}^2 = \tau B_y{}^2/\mu$, and $RT = a^2/\gamma$ to put the results in a more convenient form. Finally, using the standard procedure for solution of linear differential equations, assume all dependent variables to be of the form

$$\tau^* = \tilde{\tau}^* \exp \lambda x$$

The result of the above operations is the matrix equation 5.25, in which the subscript i has been dropped because there is no cause for confusion.

$$\begin{bmatrix}
u^2\left(1 - \frac{a^2}{\gamma u^2} - \frac{m_1}{G}\lambda\right) & 0 & a^2/\gamma & b_x b_y & 0 \\
0 & u^2\left(1 - \frac{m_2}{G}\lambda\right) & 0 & -b_x{}^2 & 0 \\
a^2/\gamma & 0 & \left(\frac{a^2}{\gamma(\gamma - 1)} - \frac{\kappa T}{G}\lambda\right) & 0 & 0 \\
b_x b_y & & 0 & b_x{}^2\left(1 - \frac{\lambda}{\mu\sigma u}\right) & -\left(\frac{m_- u^2 B_x}{e G \mu}\right)\lambda \\
0 & 0 & 0 & -\left(\frac{m_- u^2 B_x}{e G \mu}\right)\lambda & \frac{m_-}{m_+} u^2
\end{bmatrix}
\begin{vmatrix}
\bar{\tau}^* \\ \bar{v}^* \\ \bar{T}^* \\ \bar{B}_y{}^* \\ \bar{J}_z{}^*
\end{vmatrix} = 0$$

(5.25)

Inasmuch as the matrix of Equations 5.25 has no term proportional to $\lambda$ in the last diagonal element, it possesses some unusual properties, which we shall review here. This treatment begins from one given by Goldstein,[56] but is extended to emphasize properties of the present system.

Equations 5.25 can be written in the following tensor notation:

$$V_{ij} a_{jk} = \lambda_{(k)} T_{ij} a_{jk} \tag{5.26}$$

in which the parentheses indicate that the index $k$ is not summed. The quantity $a_{jk}$ is the $j^{th}$ component of the $k^{th}$ eigenvector, and it is clear, from Equations 5.25, that both $V_{ij}$ and $T_{ij}$ are symmetric. Equations 5.26 have a nontrivial solution only if their determinant vanishes. Thus, the eigenvalues can be found from

$$|V_{ij} - \lambda T_{ij}| = 0$$

and, then, each of the eigenvectors can be derived, to within a multiplicative constant, from Equation 5.26.

If we take the complex conjugate of Equation 5.26 and interchange the indices $i$ and $j$, the symmetry of $V_{ij}$ and $T_{ij}$ allows us to write

$$V_{ij} a_{i\ell}^* = \lambda_{(\ell)}^* T_{ij} a_{i\ell}^* \tag{5.27}$$

for the $\ell^{th}$ eigenvalue. Then, if we take the inner products of Equation 5.26 with $a_{i\ell}^*$ and of Equation 5.27 with $a_{jk}$ and subtract the two resulting equations, the left side vanishes and

$$0 = \left(\lambda_{(k)} - \lambda_{(\ell)}^*\right) a_{i\ell}^* T_{ij} a_{jk} \tag{5.28}$$

Consider the case $k = \ell$, in which

$$0 = \left(\lambda_{(k)} - \lambda_{(k)}^*\right) a_{ik}^* T_{ij} a_{jk} \tag{5.29}$$

**Then, letting** $a_{ij} = \alpha_{ij} + i\beta_{ij}$, we have

$$a_{ik}^* T_{ij} a_{jk} = T_{ij}[\alpha_{ik}\alpha_{jk} + \beta_{ik}\beta_{jk} + i(\alpha_{ik}\beta_{jk} - \beta_{ik}\alpha_{jk})]$$

If we reverse the indices $i$ and $j$ in the last term and take into account the symmetry of $T_{ij}$, it is evident that the imaginary part of the above expression vanishes, with the result that $T_{ij} a_{ik}^* a_{jk}$ is real. From Equation 5.25 we can write this quadratic form in the present case as

$$T_{ij}a_i^* a_j = T_{11}|a_1|^2 + T_{22}|a_2|^2 + T_{33}|a_3|^2 + T_{44}|a_4|^2 + T_{45}(a_4^* a_5 + a_4 a_5^*) \qquad (5.30)$$

in which

$$\left.\begin{aligned} T_{11} &= \frac{m_1 u^2}{G} \\ T_{22} &= \frac{m_2 u^2}{G} \\ T_{33} &= \frac{\kappa T}{G} \\ T_{44} &= \frac{b_x^2}{\mu\sigma u} \\ T_{45} = T_{54} &= \frac{m_- u^2 B_x}{eG\mu} \end{aligned}\right\} \qquad (5.31)$$

If the quadratic form of Equation 5. 30 were positive definite, as would be the case if the last row and column of Equation 5.25 were removed, Equation 5. 29 would show that $\lambda_{(k)} = \lambda_{(k)}^*$. Hence, $\lambda_{(k)}$ would have to be real. But since Equation 5.30 may be zero, complex eigenvalues are possible, and indeed they are exhibited in the following section.

The orthogonality properties of Equation 5.26 can be seen most easily by considering the equation analogous to Equation 5.28, but without having taken complex conjugates. In this case, when $k \neq \ell$, the quadratic form must vanish (if the eigenvalues are unequal); hence,

$$\tilde{a}_{\ell i} T_{ij} a_{jk} = 0 \qquad \text{if } \ell \neq k \qquad (5.32)$$

in which the tilde denotes the transpose of the tensor beneath.†

When $k = \ell$, the magnitudes of each of the five terms on the left side of Equation 5.32 are arbitrary and can be set equal to

† The use of the transpose is necessary if Equation 5.32 is to be calculated by using the rules of matrix algebra.[57]

unity; however, their signs depend upon the properties of the matrix tensor $T_{ij}$. The easiest way to see this is to find by direct computation a matrix $a_{ij}$ which will satisfy Equation 5.32, and which will give diagonal elements of unit magnitude. In the present case, the matrix $T_{ij}$ has the form

$$T_{ij} = \begin{bmatrix} T_{11} & 0 & 0 & 0 & \\ 0 & T_{22} & 0 & 0 & \\ 0 & 0 & T_{33} & 0 & \\ 0 & 0 & 0 & T_{44} & T_{45} \\ 0 & 0 & 0 & T_{45} & 0 \end{bmatrix} \qquad (5.33)$$

A little study will show that the simplest form of the matrix $a_{ij}$ is

$$a_{ij} = \begin{bmatrix} T_{11}^{-\frac{1}{2}} & 0 & 0 & 0 & 0 \\ 0 & T_{22}^{-\frac{1}{2}} & 0 & 0 & 0 \\ 0 & 0 & T_{33}^{-\frac{1}{2}} & 0 & 0 \\ 0 & 0 & 0 & T_{44}^{-\frac{1}{2}} & A \\ 0 & 0 & 0 & 0 & B \end{bmatrix}$$

Therefore,

$$\tilde{a}_{\ell i} T_{ij} a_{jk} = \begin{bmatrix} 1 & 0 & 0 & 0 & 0 \\ 0 & 1 & 0 & 0 & 0 \\ 0 & 0 & 1 & 0 & 0 \\ 0 & 0 & 0 & 1 & T_{44}^{-\frac{1}{2}}(T_{44} A + T_{45} B) \\ 0 & 0 & 0 & T_{44}^{-\frac{1}{2}}(T_{44} A + T_{45} B) & A(T_{44} A + 2 T_{45} B) \end{bmatrix}$$

(5.34)

Equation 5.32 dictates that $T_{44} A + T_{45} B = 0$; thus the fifth diagonal element becomes $-T_{44} A^2$. Referring to Equations 5.31, it is clear that this term is negative definite and obviously cannot be set equal to unity. Hence, it is normalized to $-1$, with the result that

$$A = T_{44}^{-\frac{1}{2}} \qquad B = -\frac{T_{44}^{\frac{1}{2}}}{T_{45}}$$

and $a_{ij}$ becomes

$$a_{ij} = \begin{bmatrix} T_{11}^{-\frac{1}{2}} & 0 & 0 & 0 & 0 \\ 0 & T_{22}^{-\frac{1}{2}} & 0 & 0 & 0 \\ 0 & 0 & T_{33}^{-\frac{1}{2}} & 0 & 0 \\ 0 & 0 & 0 & T_{44}^{-\frac{1}{2}} & T_{44}^{-\frac{1}{2}} \\ 0 & 0 & 0 & 0 & -\frac{T_{44}^{-\frac{1}{2}}}{T_{45}} \end{bmatrix} \tag{5.35}$$

Equations 5.34 can now be written symbolically in the form

$$\tilde{a}_{\ell i} T_{ij} a_{jk} = g_{\ell k} \tag{5.36}$$

in which $g_{\ell k}$ has the meaning

$$g_{\ell k} = \begin{bmatrix} 1 & & & & \\ & 1 & & & \\ & & 1 & & \\ & & & 1 & \\ & & & & -1 \end{bmatrix} \tag{5.37}$$

In this form, the tensor $g_{\ell k}$ is recognized as the metric tensor of the special theory of relativity, except that there are four ordinary space dimensions instead of three. Hence, the peculiar orthogonality properties of "space-time"[58] (to use relativistic terminology) fortuitously apply to the present problem. These orthogonality properties — important for the work of Section 6 — can be easily visualized as follows. Equation 5.36 is the rule for taking the inner (dot) product of two vectors. Assuming that $T_{ij}$ has been diagonalized and normalized into the form $g_{ij}$, the inner product of two vectors $\vec{a}_\ell$ and $\vec{a}_k$ is

$$\vec{a}_\ell \quad \vec{a}_k = \tilde{a}_{\ell i} g_{ij} a_{jk} = a_{1\ell} a_{1k} + \cdot \cdot \cdot + a_{4\ell} a_{4k} - a_{5\ell} a_{5k} \tag{5.38}$$

Let the two vectors lie in the direction of, say, the first and second axes. Then, expressing the components in the form

$$a_{1\ell} = a_\ell \cos\theta_\ell \qquad a_{1k} = a_k \cos\theta_k$$

$$a_{2\ell} = a_\ell \sin\theta_\ell \qquad a_{2k} = a_k \sin\theta_k$$

in which $a_{\ell,k}$ and $\theta_{\ell,k}$ are the magnitudes and inclinations of $\vec{a}_{\ell,k}$ from the first axis, one obtains the inner product

$$\vec{a}_\ell \cdot \vec{a}_k = a_\ell a_k \cos(\theta_\ell - \theta_k) \tag{5.39}$$

But suppose the two vectors lie, say, in the direction of the fourth and fifth axes. Then, because of the minus sign in front of the last term in Equation 5.38, the inner product becomes

$$\vec{a}_\ell \cdot \vec{a}_k = a_\ell a_k \cos(\theta_\ell + \theta_k) \tag{5.40}$$

in which the amplitudes and angles have meanings analogous to those above.

Comparison of Equation 5. 39 with Equation 5. 40 reveals the profound difference between the orthogonality properties of vectors in ordinary space with those of vectors in "space-time." Two vectors are said to be orthogonal when their inner product is zero. Thus, Equation 5. 39 shows that in ordinary space two vectors are orthogonal when they are 90 degrees apart, and no direction is preferred. In space-time (or when one component of each of the two vectors lies along the fifth axis) the orthogonality property, $\theta_\ell + \theta_k = \pm 90$ degrees, is decidedly anisotropic For example, if $\vec{a}_\ell$ is along the positive time axis ($\theta_\ell = 90$ degrees) and $\vec{a}_k$ is along the positive space axis ($\theta_k = 0$), the two vectors are orthogonal and also 90 degrees apart; however, they are also orthogonal if, say, $\theta_\ell = \theta_k = 45$ degrees, in which case they are colinear. Recognition of this peculiar property will be useful later in this chapter.

The matrix $a_{ij}$ has been shown (see Equation 5. 36) to diagonalize $T_{ij}$ by what Goldstein[56] calls a "congruent transformation," the general form of which is $T' = \tilde{A}TA$. Using Equation 5.26, we can see as follows that it also diagonalizes $V_{ij}$ by the same congruent transformation. Accordingly, premultiply both sides of Equation 5.26 by $\tilde{a}_{\ell i}$. Then, use Equation 5.36 to obtain

$$\tilde{a}_{\ell i} V_{ij} a_{jk} = \lambda_{(k)} g_{\ell k}$$

a diagonal form, from which

$$\lambda_{(k)} = \tilde{a}_{ki} V_{ij} a_{jk} \qquad (k = 1, \quad \cdot, 4)$$

and

$$\lambda_5 = - \tilde{a}_{5i} V_{ij} a_{j5}$$

These expressions would give the eigenvalues if the original axes had been chosen in the direction of the eigenvectors, i.e., if the original system had been uncoupled. The eigenvectors are, of course, unknown at the beginning of the problem and must be found by the usual procedure, outlined below Equation 5.26. The above theory does, however, show the angular relationship between the eigenvectors, which will be of considerable help in the following sections. Thus, as indicated below Equation 5.26, $a_{ij}$ is the $i^{th}$ component of the $j^{th}$ eigenvector. Each column of Equation 5.35 yields the components of one of the five eigenvectors with respect to a set of orthogonal principal coordinates — four spacelike and one timelike. Hence, the first four eigenvectors form a set of four-dimensional Cartesian coordinates and are orthogonal to each other in the ordinary sense. The fifth eigenvector has one component along the fifth or timelike axis, and one component along the fourth spacelike axis.

Thus, the fifth eigenvector is skewed out of orthogonality (in the ordinary sense) in the direction of the fourth principal axis. The degree of skew can be seen by taking the ratio of the two components, $T_{44}/T_{45}$. From Equations 5.31,

$$\frac{T_{44}}{T_{45}} = \frac{b_x^2}{u^2} \frac{e^2 n}{m_- \sigma} \frac{m_-}{eB_x} \frac{m_+}{m_-} = \left(\frac{b_x}{u}\right)^2 \left(\frac{\nu_c}{\omega_c}\right) \frac{m_+}{m_-} \gg 1$$

in which the definitions of collision frequency $\nu_c$ and cyclotron frequency $\omega_c$ have been taken from Equations 4.62 and 4.63. The first factor is of the order unity, the second factor is large according to a basic assumption of this work, and the third factor is always large. Hence, the skew in the fifth eigenvector is very small and, in the following analysis, can be considered to lie in the negative direction of the timelike principal axis.

The orthogonality property represented by Equation 5.32 can be applied directly if the eigenvalues are real; but if there is a pair of complex-conjugate eigenvalues, the corresponding complex-conjugate eigenvectors do not correspond to real directions in space. In this case, the singular point will be a spiral point in a plane which can be described as being orthogonal to the three-space of the three real eigenvectors; however, a more direct method of locating the plane of the spiral will be needed for the existence proofs of this chapter. The basis for this method will now be given.

In Equation 5.28, let $\lambda_{(k)}$ be real and $\lambda_{(\ell)}$ complex. Then let

$$\lambda_{(\ell)} = \alpha_{(\ell)} + i\beta_{(\ell)}$$

$$a_{i\ell} = A_{i\ell} + i\,B_{i\ell}$$

After substituting these definitions into Equation 5.28, expand the result and set the real and imaginary parts equal to zero separately. The result is

$$0 = (\lambda_{(k)} - \alpha_{(\ell)})\,\tilde{A}_{\ell i} T_{ij} a_{jk} + \beta_{(\ell)} \tilde{B}_{\ell i} T_{ij} a_{jk}$$

$$0 = \beta_{(\ell)} \tilde{A}_{\ell i} T_{ij} a_{jk} - (\lambda_{(k)} - \alpha_{(\ell)})\,\tilde{B}_{\ell i} T_{ij} a_{jk}$$

The determinant of these equations,

$$-(\lambda_{(k)} - \alpha_{(\ell)})^2 - \beta_{(\ell)}^{\;2}$$

is negative definite, since $\lambda_{(k)}$ is assumed to be different from $\lambda_{(\ell)}$; hence,

$$\left.\begin{aligned} \tilde{A}_{\ell i}\, T_{ij}\, a_{jk} &= 0 \\ \tilde{B}_{\ell i}\, T_{ij}\, a_{jk} &= 0 \end{aligned}\right\} \qquad (5.41)$$

This shows that the real part and imaginary part of the complex eigenvector each forms a separate vector which is orthogonal, in the sense of Equation 5.32, to the three real eigenvectors of the five-dimensional system. Since $A_{i\ell} \neq B_{i\ell}$ in general, these two new vectors (usually not orthogonal to each other) define the plane of the spiral, and in a way which will be of direct benefit later in this chapter.

The relationship between the original set of coordinates $\tau$, $v$, $T$, $B_y$, $J_z$ and the principal coordinates remains to be clarified. In this discussion, if there is a set of complex-conjugate eigenvectors, a pair of real axes is chosen in their plane as indicated above; hence, the concept of a five-dimensional orthogonal set of axes can be retained. It is necessary to realize that in doing geometry in a configuration space of mathematical dependent variables, one can choose coordinates in any convenient manner.

With this in mind, choose the $\tau$-v-T-$B_y$-$J_z$ axes as ordinary Cartesian coordinates. Call this space $S_c$. (One can visualize $S_c$ a three-space at a time.) Then calculate the five components of each of the five eigenvectors from Equations 5.25. After doing this, one knows the orientation of each of the eigenvectors, which will be orthogonal in the sense of Equations 5.32. Thus, from the geometric properties described above, four of the eigenvectors will be orthogonal to each other in the usual sense, but the orientation in $S_c$ of the fifth (timelike) eigenvector depends on the orientation of the four spacelike eigenvectors with four of the axes of $S_c$. Consequently, it can be almost anywhere.

## 5. Analysis of the Eigenvalues

In this section we give a qualitative analysis of the eigenvalues of Equations 5. 25 to determine their signs, and the conditions under which they are complex.

Let $D_5$ represent the determinant of Equations 5.25. Then the eigenvalues are found from the polynomial $D_5(\lambda) = 0$. Expanding by minors, we can write this equation in the form

$$D_4'(\lambda) - \frac{m_+ m_-}{e^2} \frac{b_x^{\,2}\tau}{\mu} \lambda^2 D_3'(\lambda) = 0 \tag{5.42}$$

in which $D_4'$ ($D_3'$) is the determinant of the first four (three) rows and columns of $D_5$.

More meaningful results can be obtained by using the dimensionless eigenvalues

$$\Lambda = \lambda L \tag{5.43}$$

and by dividing through each of the determinants $D_4'$ and $D_3'$ by the coefficient of the highest power of $\Lambda$. Then two new determinants, $D_4(\Lambda)$ and $D_3(\Lambda)$, are obtained in which

$$\frac{D_4'(\lambda)}{D_3'(\lambda)} = -\frac{b_x^{\,2}}{\mu\sigma u L} \frac{D_4(\Lambda)}{D_3(\Lambda)}$$

Equation 5.42 then becomes

$$D_4(\Lambda) + \eta\, \Lambda^2 D_3(\Lambda) = 0 \tag{5.44}$$

in which

$$\eta = \frac{u}{L\nu_c} \tag{5.45}$$

Thus, if $L$ is of the order of the shock thickness, it will be expected (from Section 3 of Chapter 4) that $\eta$ is of order unity.

The positive definite quantity $\eta$ will be treated as a variable parameter. When $\eta = 0$, four of the eigenvalues of $D_5 = 0$ are those of $D_4 = 0$ (Germain's solution[13]), and the fifth is off at infinity. Then, as $\eta$ increases — and with it the effect of electron inertia — the eigenvalues vary in a manner that will be discussed in detail.

The first step is to determine the signs of the eigenvalues of $D_4$ and $D_3$. Following Germain,[13] this is easily accomplished with the help of the "Inertial Theorem for Quadratic Forms":[59]

> The number of positive and negative coefficients, respectively, in a quadratic form reduced to an expression $\Sigma c_p z_p^2$ by means of a nonsingular real linear transformation does not depend on the particular transformation.

But among the family of nonsingular real linear transformations, there is one which will diagonalize the matrix of Equations 5.25, in which case the $c_p$ in the above quadratic form are the eigenvalues. Using this knowledge, the inertial theorem shows that the number of positive and negative eigenvalues of $D_4$ and $D_3$ can be found in the following way: As shown in the preceding section, equations like those of 5.25 can be written, for a particular eigenvalue, in the form

$$(A_{ij} - \lambda T_{ij}) a_j = 0$$

in which $a_j$ is the eigenvector. Quadratic forms are obtained by taking the inner product of this expression with $a_i$. Then,

$$A_{ij} a_i a_j = \lambda T_{ij} a_i a_j \tag{5.46}$$

For the system derived by striking out the last row and column of Equations 5.25,

$$T_{ij} a_i a_j = \frac{m_1 u^2}{G} \bar{\tau}^{*2} + \frac{m_2 u^2}{G} \bar{v}^{*2} + \frac{\kappa T}{G} \bar{T}^{*2} + \frac{b_x^2}{\mu \sigma u} \bar{B}_y^{*2}$$

and

$$\begin{aligned} A_{ij} a_i a_j = {} & \left(u^2 - \frac{a^2}{\gamma}\right) \bar{\tau}^{*2} + 2\frac{a^2}{\gamma} \bar{\tau}^* \bar{T}^* + 2b_y b_x \bar{\tau}^* \bar{B}_y^* + u^2 \bar{v}^{*2} - 2b_x^2 \bar{v}^* \bar{B}_y^* \\ & + \frac{a^2}{\gamma(\gamma-1)} \bar{T}^{*2} + b_x^2 \bar{B}_y^{*2} = \left(b_x \bar{B}_y^* + b_y \bar{\tau}^* - b_x \bar{v}^*\right)^2 \\ & + \frac{a^2}{\gamma(\gamma-1)} \left(\bar{T}^* + (\gamma - 1) \bar{\tau}^*\right)^2 + (u^2 - b_x^2)\left(\bar{v}^* + \frac{b_x b_y}{u^2 - b_x^2} \bar{\tau}^*\right)^2 \\ & + \left(u^2 - a^2 - b_y^2 - \frac{b_x^2 b_y^2}{u^2 - b_x^2}\right) \bar{\tau}^{*2} \end{aligned} \tag{5.47}$$

By making an obvious linear transformation of variables, one can convert the form of Equation 5.47 into the form $\Sigma c_p z_p^2$ as required by the inertial theorem. Then, when Equation 5.47 is subjected to a further transformation to bring the whole system of Equations 5.46 into diagonal form, the inertial theorem states that the number of positive and negative coefficients $c_p$ does not change. Hence, the number of positive and negative coefficients in front of the squared factors in Equation 5.47 is the same as the number of positive and negative eigenvalues.

The first two of these coefficients are clearly positive definite. The third is positive when the x-component of the flow velocity exceeds the x-component of the Alfvén velocity or, in terms of the theory of Chapter 2, it is positive at $SP_1$ and $SP_2$ (Figure 2.5) and negative at $SP_3$ and $SP_4$. (Again, $SP_i$ refers to the $i^{th}$ singular point.) The behavior of the fourth coefficient is more easily understood when expressed, with the help of Equation 2.19, in the form

$$\frac{u^4 - (a^2 + b_x^2 + b_y^2)u^2 + a^2 b_x^2}{u^2 - b_x^2} = \frac{(u^2 - c_f^2)(u^2 - c_s^2)}{u^2 - b_x^2}$$

Thus, this coefficient is positive at $SP_1$ and $SP_3$ and negative at $SP_2$ and $SP_4$. Summarizing, the eigenvalues of $D_4$ are oriented as follows:

$$\left.\begin{array}{ll} SP_1 : & 0 < \lambda_1 < \lambda_2 < \lambda_3 < \lambda_4 \\ SP_2 : & \lambda_1 < 0 < \lambda_2 < \lambda_3 < \lambda_4 \\ SP_3 : & \lambda_1 < 0 < \lambda_2 < \lambda_3 < \lambda_4 \\ SP_4 : & \lambda_1 < \lambda_2 < 0 < \lambda_3 < \lambda_4 \end{array}\right\} \qquad (5.48)$$

Similarly, we can find the eigenvalues of $D_3$ by considering the quadratic form obtained by striking out the last two rows and columns of 5.25. Then, for this case,

$$A_{ij}a_i a_j = \frac{a^2}{\gamma(\gamma - 1)}(\bar{T}^* + (\gamma - 1)\bar{\tau}^*)^2 + u^2 \bar{v}^{*2} + (u^2 - a^2)\bar{\tau}^{*2}$$

Consequently, the eigenvalues of $D_3$ are all positive if $u^2 > a^2$, and there are two positive and one negative eigenvalue if $u^2 < a^2$ Referring to the Inequality 2.51, and taking into account that the relationship between $u^2$ and $a^2$ at $SP_2$ and $SP_3$ is not definite, we can arrange the eigenvalues $\bar{\lambda}$ of $D_3$ as follows:

$$\left.\begin{array}{ll} SP_1 : & 0 < \bar{\lambda}_1 < \bar{\lambda}_2 < \bar{\lambda}_3 \\ SP_{2,3}\,(u > a): & 0 < \bar{\lambda}_1 < \bar{\lambda}_2 < \bar{\lambda}_3 \\ SP_{2,3}\,(u < a): & \bar{\lambda}_1 < 0 < \bar{\lambda}_2 < \bar{\lambda}_3 \\ SP_4 : & \bar{\lambda}_1 < 0 < \bar{\lambda}_2 < \bar{\lambda}_3 \end{array}\right\} \quad (5.49)$$

From the theory in Section 4 of Chapter 2, it is clear that for fast shocks, $u > a$ at $SP_2$ if the shock is weak, but the inequality will eventually reverse itself as the shock strength increases if $a_2 > b_{x_2}$. For slow shocks, the opposite condition exists at $SP_3$.

The next step is to determine the relation between the eigenvalues of $D_3$ and $D_4$ at a specific singular point. This relation is

$$\lambda_1 < \bar{\lambda}_1 < \lambda_2 < \bar{\lambda}_2 < \lambda_3 < \bar{\lambda}_3 < \lambda_4 \quad (5.50)$$

i.e., the eigenvalues of $D_4$ bracket those of $D_3$. There are at least two completely different ways to prove this. One method is given by Courant and Hilbert[59] and results from a problem in which each eigenvalue is characterized as the minimum of the maximum (or vice versa) of a quadratic form.

The second and more straightforward method depends on an interesting theorem from the theory of determinants,[60] the proof of which is found in many texts on determinants written from 50 to 100 years ago. For symmetric determinants, this theorem can be specialized to

$$\frac{\partial D}{\partial a_{11}} \frac{\partial D}{\partial a_{22}} - \left(\frac{\partial D}{\partial a_{12}}\right)^2 = D \frac{\partial^2 D}{\partial a_{11} \partial a_{22}} \quad (5.51)$$

in which $D = \det(a_{ij})$. By the theorem on expansion of determinants by minors, $\partial D/\partial a_{ij}$ is the cofactor (signed minor) of the term $a_{ij}$; hence, if $D = D_4$, the indices can be arranged so that $\partial D/\partial a_{11} = D_3$ and $\partial^2 D/\partial a_{11} \partial a_{22} = D_2$, where $D_2$ is the determinant obtained from Equations 5.25 by striking out the last three rows and columns. Then at each zero of $D_3$, Equation 5.33 becomes

$$-\left(\frac{\partial D}{\partial a_{12}}\right)^2 = D_4 \, D_2$$

i.e., $D_4$ and $D_2$ have opposite signs. Furthermore, it is clear that this relationship exists for all sets of subdeterminants $D_{j-1}$, $D_j$, $D_{j+1}$ of $D_n$. All of these relationships can hold only if the eigenvalues (zero points) of each successively higher-order determinant bracket the eigenvalues of the next-lower determinant.

Using the theory developed above, Figure 5.6 depicts $\lambda^2 D_3(\lambda)$, $D_4(\lambda)$, and, from Equation 5.44, $D_4(\lambda) + \eta \lambda^2 D_3(\lambda)$ as they appear at $SP_1$. The arrows indicate how the roots of $D_5$ move away

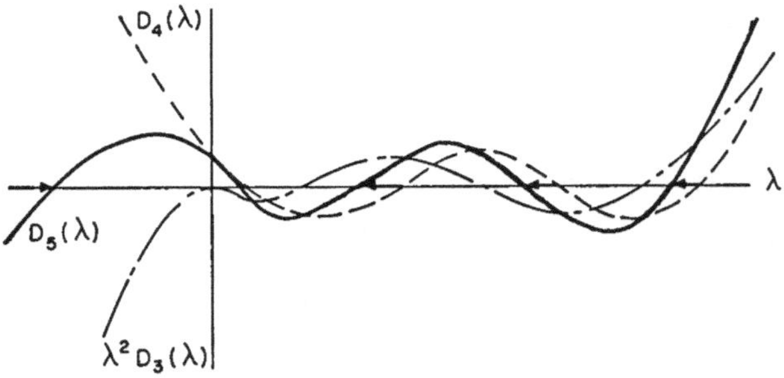


Figure 5.6. Location of the eigenvalues of the fifth-order linearized system at $SP_1$ (ahead of a fast shock

from the roots of $D_4$ as $\eta$ increases. Note that as $\eta$ ranges from zero to infinity, the largest root of $D_5$ moves from the largest root of $D_4$ down to the largest root of $D_3$. Similarly, the second and third largest roots of $D_5$ are restricted to identical ranges with respect to the second and third largest roots of $D_3$ and $D_4$. The fourth largest root of $D_5$ ranges from the lowest root of $D_4$ down to zero, and the leftmost root of $D_5$ (the only one that moves to the right as $\eta$ increases) is always negative and moves from $-\infty$ up to zero. Summarizing, we can say that at $SP_1$ the $\lambda$'s are all real for all $\eta > 0$, and they can be ordered as follows:

$$\lambda_1 < 0 < \lambda_2 < \lambda_3 < \lambda_4 < \lambda_5 \tag{5.52}$$

The same set of functions of $\lambda$ are shown in Figure 5.7 for a small value of $\eta$ when $u > a$ at $SP_2$ or $SP_3$. The three largest roots of $D_5$ behave the same way as at $SP_1$; however, the two smallest roots, both negative when they are real, move towards each other and meet for some finite value of $\eta > 0$. With a further increase in $\eta$, these two roots become complex conjugates and remain so until $\eta$ equals infinity. As $\eta$ approaches infinity, $D_4$ in Equation 5.44 becomes negligible compared with $\eta \lambda^2 D_3$; hence, the pair of complex conjugate roots approaches the origin. Again summarizing, we see that when the $\lambda$'s at $SP_2$ or $SP_3$ with $u > a$ are real, they are ordered as follows:

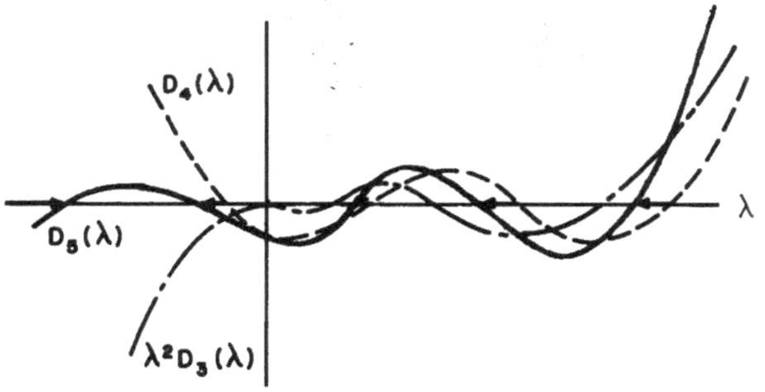


Figure 5.7. Location of the eigenvalues of the fifth-order linearized system at $SP_2$ or $SP_3$ when $u > a$

$$\lambda_1 < \lambda_2 < 0 < \lambda_3 < \lambda_4 < \lambda_5 \tag{5.53}$$

The smallest pair, $\lambda_1$ and $\lambda_2$, may be complex, but it is not yet known whether $Re(\lambda_1, \lambda_2)$ remain negative as $\eta$ increases.

When $u < a$ at $SP_2$ or $SP_3$, a more peculiar behavior results. The same set of functions for this case is illustrated in Figure 5.8 for a small value of $\eta$. The difference in the behavior of this

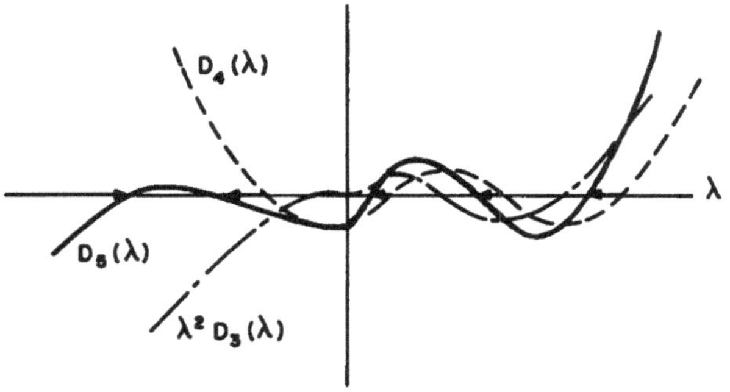


Figure 5.8. Location of the eigenvalues of the fifth-order linear system at $SP_2$ or $SP_3$ when $u < a$

case compared with the preceding one (Figure 5.7) is that as $\eta$ increases, the smallest pair of eigenvalues are first real, then complex, then real again. For large $\eta$ they reappear between the lowest $\lambda$ of $D_3$ and zero, and then as $\eta$ approaches infinity they approach these two points. When they are real, Inequality 5.53 still holds.

At $SP_4$, Figure 5.9 describes the behavior of the eigenvalues. Here the general behavior is similar to that shown in Figure 5.7, and when the eigenvalues are real, they are ordered as follows:

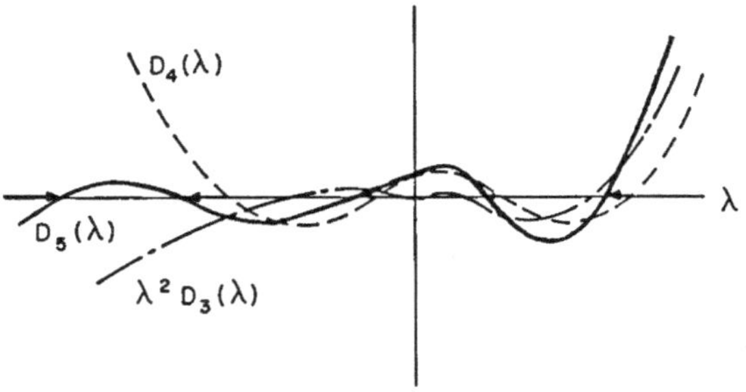


Figure 5.9. Location of the eigenvalues of the fifth-order linearized system at $SP_4$ (behind the slow shock)

$$\lambda_1 < \lambda_2 < \lambda_3 < 0 < \lambda_4 < \lambda_5 \tag{5.54}$$

The real parts of the complex eigenvalues described above are negative for the $\eta$ for which they just leave the real axis. We shall now prove that they remain negative for all $\eta > 0$.

Consider the complex $\lambda$-plane. Represent $\lambda$ in polar form by

$$\lambda = Re^{i\theta} \tag{5.55}$$

and the polynomials $D_4(\lambda)$ and $D_3(\lambda)$ in the form

$$D_4(\lambda) = \prod_1^4 (\lambda - \lambda_i) = \prod_1^4 r_i e^{i\phi_i} = re^{i\phi} \tag{5.56}$$

$$D_3(\lambda) = \prod_1^3 (\lambda - \bar{\lambda}_i) = \rho e^{i\psi} \tag{5.57}$$

where

$$r = \prod_1^4 r_i \quad , \quad \phi = \sum_1^4 \phi_i$$

$$\rho = \prod_1^3 \rho_i \quad \quad \psi = \sum_1^3 \psi_i$$

In the $\lambda$-plane, each $\lambda - \lambda_i$ is a vector from $\lambda_i$ to $\lambda$, as shown in Figure 5.10, in which the position of the origin is correct for the case corresponding to Figure 5.7.

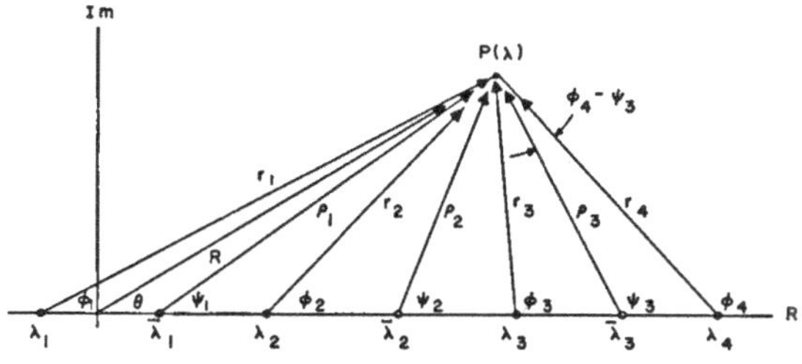


Figure 5.10. A point in the complex λ-plane in relation to the eigenvalues of $D_4(\lambda)$ and $D_3(\lambda)$

Substituting Equations 5.55, 5.56, and 5.57 into Equation 5.44 yields

$$re^{i\phi} = \eta R^2 \rho e^{i(2\theta + \psi + \pi)}$$

which can be satisfied only if the magnitudes of the two sides are equal, and if their arguments are equal within a factor of $2n\pi$. Hence,

$$\Phi \equiv 2\theta + \psi + \pi - \phi = 0,\ 2\pi,\ 4\pi,\ \cdots \qquad (5.58)$$

at all points in the λ-plane at which Equation 5.44 is satisfied, and it is important to notice that this locus of roots does not depend on the parameter $\eta$. Thus, as $\eta$ varies, the eigenvalues of $D_5$ move along a fixed locus defined by Equation 5.58.

It will be shown that this locus cannot cross the imaginary axis, but first it is useful to verify from Equation 5.58 that the real roots of $D_5$ are as shown in Figure 5.7. Write $\Phi$ in the form

$$\Phi = (\pi - \phi_4) + (\psi_3 - \phi_3) + (\psi_2 \quad \phi_2) + (\psi_1 - \phi_1) + 2\theta$$

Since each of the terms in parentheses is positive at every point in the upper half plane, $\Phi$ is positive at every point in the upper half plane. At a point on the real axis where $\lambda > \lambda_4$, $\Phi = \pi$; therefore, no roots are possible there. When $\bar{\lambda}_3 < \lambda < \lambda_4$, $\Phi = 0$; hence, there is a root in this region. But for all points an infinitesimal distance above the real axis in the interval $\bar{\lambda}_3 < \lambda < \lambda_4$, $\Phi > 0$; so there can be no roots there. In this way it is verified that no complex roots can leave the real axis in this region. By repeating this analysis, it becomes obvious that real roots are possible only within the regions of the real axis where they are indicated in Figure 5.7.

Now let $0 = \pi/2$; that is, consider the upper half of the imaginary axis. Then Equation 5.58 becomes

$$\Phi = 2\pi + \psi - \phi = 2\pi - (\phi_4 - \psi_3) - (\phi_3 - \psi_2) - (\phi_2 - \psi_1) - \phi_1 \quad (5.59)$$

Now note, from Figure 5.10, that points $\lambda_4$ and $\bar{\lambda}_3$, say, subtend an angle $\phi_4 - \psi_3$ at point P, and that each vertex angle at P has a similar meaning. This means that as $\lambda$ moves along the positive imaginary axis in the region bounded away from zero and $i\infty$,

$$0 < (\phi_4 - \psi_3) + (\phi_3 - \psi_2) + (\phi_2 - \psi_1) < \pi$$

and

$$0 < \phi_1 < \frac{\pi}{2}$$

whenever the origin of the $\lambda$-plane lies between $\lambda_1$ and $\lambda_4$. Therefore, from Equation 5.59,

$$\frac{\pi}{2} < \Phi < 2\pi$$

Also, $\Phi \to 3\pi/2$ as $\lambda \to i\infty$, but $\Phi \to 2\pi$ as $\lambda \to 0$. Thus, the origin is the only point on the imaginary axis at which $\Phi$ satisfies Equation 5.58, and it has been already indicated that $\lambda = 0$ is a root of $D_5 = 0$ only for $\eta = \infty$. Combining this with the above proof that the root locus cannot leave the real axis on the positive side, we have now proved that the real parts of the complex eigenvalues remain negative for all finite $\eta$. It is also clear that the above proof is valid for any location of the origin of the $\lambda$-plane between $\lambda_1$ and $\lambda_4$. Thus, this proof is valid for all cases of interest. QED

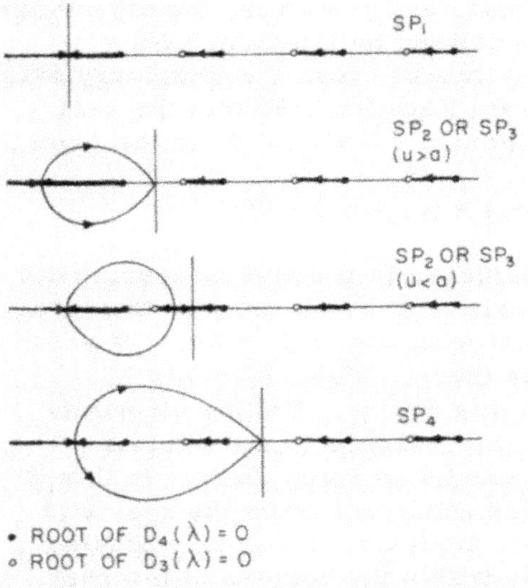


Figure 5.11. The loci, in the complex $\lambda$-plane, of the roots of $D_5(\lambda) = 0$. Arrows indicate the progress of the roots as $\eta$ goes from 0 to $+\infty$

The results of this study of eigenvalues can be summarized by plotting, in the complex $\lambda$-plane, the loci of the roots of $D_5(\lambda)$ as $\eta$ goes from zero to infinity. These loci are shown in Figure 5.11.

## 6. Location of the Eigenvectors

In Section 3, we mentioned that five null surfaces separate the five-dimensional configuration space into various regions. The next section will show that in the case when all four singular points are present, there are 52 such regions adjacent to the singular points. The task of the present section is to find, as far as possible, which of these regions contain eigenvectors.† Once we know the signs of the eigenvectors, we can obtain the means for doing this very simply from Equations 5.13 through 5.17, written in the form

$$m_1 G \frac{d\tau}{dx} = F_\tau(\tau, T, B_y) \tag{5.60}$$

$$m_2 \frac{dv}{dx} = F_v(v, B_y) \tag{5.61}$$

$$\frac{\kappa}{G} \frac{dT}{dx} = F_T(\tau, T, B_y, v) \tag{5.62}$$

$$\frac{1}{\mu} \frac{dB_y}{dx} = F_B(J_z) \tag{5.63}$$

$$\nu G \tau^2 \frac{dJ_z}{dx} = F_J(\tau, T, B_y, v, J_z) \tag{5.64}$$

By equating each of the right sides to zero, we obtain the five null surfaces.

Near a singular point, Equation 5.60, for example, becomes

$$m_{1i} G \tau_i \lambda \bar{\tau}^* = dF_\tau \tag{5.65}$$

and, of course, each of the other four equations can be expressed in similar form. Furthermore, since the eigenvectors point in the direction of increasing $x$, $\bar{\tau}^* > 0$ corresponds to $d\tau = \tau(x_2) - \tau(x_1) > 0$ if $x_2 - x_1 = dx > 0$. Then, Equation 5.65 shows that $dF_\tau = F_\tau(x_2) - F_\tau(x_1) > 0$ when $\lambda\bar{\tau}^* > 0$. Consider a point leaving a singular point as $x$ increases ($\lambda > 0$). But

† The main part of this discussion applies directly if the eigenvalues are real. The additional difficulties which arise if they are complex are discussed at the end of this section.

$x_1 = -\infty$ at the singular point, so that $F_\tau(x_1) = 0$. Then if $\bar{\tau}^* > 0$, $F_\tau(x_2) > 0$; that is, the point moves into the region where $F_\tau > 0$. On the other hand, consider a point approaching a singular point as x increases ($\lambda < 0$). Let $x_2 = +\infty$ at the singular point, so that $F_\tau(x_2) = 0$. Then, if $\lambda\bar{\tau}^* > 0$, $-F_\tau(x_1) > 0$; that is, the point comes from the region in which $F_\tau < 0$. From considerations like these, the region formed by the known null surfaces in which any particular eigenvector lies can be found.

Knowledge of the signs of all five components of all five eigenvectors would completely fulfill the purpose of this section. Unfortunately, it is not possible to find all of this information; however, enough can be obtained to be of considerable help in the existence proofs.

We can find the relative signs of the components of several of the eigenvectors from Equations 5.25. By inspection, we can glean the following information from the second, third, and fifth equations and by combining the first and third equations:

$v^*$ $\bar{B}_y^*$ have same sign if $\lambda < \lambda_{m_2}$ (5.66)

$\bar{\tau}^*$ $\bar{T}^*$ have same sign if $\lambda > \lambda_\kappa$ (5.67)

$\bar{J}_z^*$ $\bar{B}_y^*$ have same sign if $\lambda > 0$ (5.68)

$\bar{\tau}^*$, $\bar{B}_y^*$ have same sign if $B_y(\lambda - \lambda_\kappa)$, $P(\lambda) > 0$

(5.69)

$\bar{T}^*$, $\bar{B}_y^*$ have same sign if $B_y P(\lambda) > 0$ (5.70)

in which

$$\lambda_{m_2} = \frac{G}{m_2} \tag{5.71}$$

$$\lambda_\kappa = \frac{a^2}{\gamma(\gamma - 1)} \frac{G}{\kappa T} \tag{5.72}$$

$$P(\lambda) = \begin{vmatrix} u^2\left(1 - \dfrac{a^2}{\gamma u^2} - \dfrac{m_1\lambda}{G}\right) & \dfrac{a^2}{\gamma} \\ \dfrac{a^2}{\gamma} & \dfrac{\kappa T}{G}(\lambda_\kappa - \lambda) \end{vmatrix} \tag{5.73}$$

To apply Conditions 5.69 and 5.70, it is necessary to know the sign of $B_y$. This is a matter of convention but, once chosen, must be adhered to consistently. In Section 2 of Chapter 2, we decided to take $\overline{F}_y > 0$; then $B_y > 0$ when $\tau > \tau^*$, that is, at $SP_1$ and $SP_2$ (see Figure 2.5), and $B_y < 0$ at $SP_3$ and $SP_4$.

Section 5 of the present chapter showed that the roots of $D_j$ separate the roots of $D_j + 1$, where $j = 1, 2, \cdots$. Applying this theorem to the subdeterminants of Equations 5.25, we can see that

$$\lambda_1, \lambda_2 < \min \{\lambda_{m_1}, \lambda_{m_2}, \lambda_\kappa, \lambda_\sigma\} \tag{5.74}$$

$$\lambda_5 > \max \{\lambda_{m_1}, \lambda_{m_2}, \lambda_\kappa, \lambda_\sigma\} \tag{5.75}$$

$$P(\lambda_1), \; P(\lambda_2), \; P(\lambda_5) > 0 \tag{5.76}$$

where

$$\lambda_{m_1} = \frac{G}{m_1}\left(1 - \frac{a^2}{\gamma u^2}\right) \tag{5.77}$$

$$\lambda_\sigma = \mu\sigma u \tag{5.78}$$

and $\lambda_1 < \lambda_2 < \lambda_3 < \lambda_4 < \lambda_5$ are the five roots of $D_5(\lambda) = 0$.

With the help of Expressions 5.66 through 5.78, it can be seen that the signs of the eigenvector components are as given in Table 5.1. The sign of $\overline{T}^*$ has been arbitrarily chosen as positive. If

**Table 5.1. Signs of Components of Eigenvectors**

| | | $\overline{T}^*$ | $\overline{\tau}^*$ | $\overline{v}^*$ | $\overline{B}_y^*$ | $\overline{J}_z^*$ |
|---|---|---|---|---|---|---|
| $\lambda_1$ | $SP_{1,2}$ | + | - | + | + | - |
| | $SP_{3,4}$ | + | - | - | - | + |
| $\lambda_2$ | $SP_1$ | + | - | + | + | + |
| | $SP_2$ | + | - | + | + | - |
| | $SP_{3,4}$ | + | - | - | - | + |
| $\lambda_3$ | $SP_4$ | + | - | + | + | - |
| $\lambda_5$ | $SP_{1,2}$ | + | + | - | + | + |
| | $SP_{3,4}$ | + | + | + | - | - |

it were negative, then all of the signs in the table would be reversed. The signs of the components of $\vec{\lambda}_3$, the eigenvector corresponding to $\lambda_3$, have at $SP_4$ been determined partly by orthogonality requirements. They are as given in the table if $P(\lambda_3) < 0$. If $P(\lambda_3) > 0$, the signs of the three components on the right are reversed, but then all of the signs of this eigenvector agree with the signs of $\vec{\lambda}_1$ and $\vec{\lambda}_2$ at $SP_4$. As explained in Section 4, the signs of the components of $\vec{\lambda}_1$ can all agree with those of one other eigenvector at a given singular point, but not more than one. Thus the signs must be as shown.

Unfortunately, there is not enough information available to determine the signs of $\vec{\lambda}_3$ at the other three singular points, or of $\vec{\lambda}_4$ at any of the singular points. The reason for this is that these latter signs depend upon the relative values of certain of the dissipation coefficients, and hence if there are shock layers corresponding to these eigenvectors for certain values of the dissipation coefficients, they may not exist for all values. More will be said about this in connection with the discussion of slow shocks.

The determination of the directions of the eigenvectors given in Table 5.1 has been made under the assumption that $\lambda_1$ and $\lambda_2$ are real. When they are complex, the corresponding eigenvectors are also complex, and, as already mentioned, the singular point is a spiral in a plane orthogonal to the three-space of $\vec{\lambda}_3$, $\vec{\lambda}_4$, and $\vec{\lambda}_5$. We shall now show that it is possible to locate the direction of one vector in the plane of the spiral in the manner accomplished above for real eigenvectors.

This analysis is based upon Equations 5.41, which show that the plane of the spiral can be defined by two vectors, one of which is formed from the real parts of the components of the complex-conjugate eigenvectors and the other from the imaginary parts. From Equations 5.25 and 5.66 through 5.70, it can be seen that four equations of the type

$$f(\lambda)\,\bar{x} = \bar{B}_y^*$$

have been used to find the relative signs of the eigenvector components. The quantity $\bar{x}$ represents any one of the other four components, and at a particular singular point, $f(\lambda)$ is a function of $\lambda$ only. Thus, consider expressions like

$$f(\lambda)\,x = y$$

If each of these quantities is complex, this expression can be expanded into

$$(f_R + if_I)(x_R + ix_I) = y_R + iy_I$$

which gives

$$f_R x_R - f_I x_I = y_R$$

$$f_I x_R + f_R x_I = y_I$$

from which

$$x_R = \frac{f_R y_R + f_I y_I}{f_R^2 + f_I^2}$$

$$x_I = \frac{-f_I y_R + f_R y_I}{f_R^2 + f_I^2}$$

But $x_R$ and $y_R$ are two components of a five-vector, say $\vec{\lambda}_R$, and $x_I$ and $y_I$ two components of another, say $\vec{\lambda}_I$, the two together forming the plane of the spiral. Hence, it is clear that the vector $\vec{\lambda}_R + c\vec{\lambda}_I$, where $c$ is a real number, is also in the plane of the spiral. Thus, from the above expressions,

$$x_R + cx_I = \frac{f_R(y_R + cy_I) + f_I(y_I - cy_R)}{f_R^2 + f_I^2}$$

The constant $c$ is arbitrary, and can be chosen so that $y_I = cy_R$. Then

$$x_R + cx_I = \frac{f_R}{f_R^2 + f_I^2}(y_R + cy_I)$$

Therefore, the sign of $x_R + cx_I$ relative to $y_R + cy_I$ depends only upon $f_R(\lambda)$, and so a vector in the plane of the spiral with easily analyzable components has been found.

In the cases of Expressions 5.66 and 5.68, $f(\lambda)$ is linear; hence, if we take the real part, the criteria for the relative signs of the components in these two expressions are the same as before. Since $\lambda_{m_2} > 0$ and $\mathrm{Re}(\lambda_1, \lambda_2) < 0$, the relative signs of $\bar{B}_y^*$, $\bar{v}^*$, and $\bar{J}_z^*$ are as in Table 5.1 for complex as well as real eigenvalues. The criteria of Equations 5.69 and 5.70 depend upon the quadratic factor $P(\lambda)$. For real $\lambda_1$ and $\lambda_2$, this factor was always positive; however, $\mathrm{Re}P(\lambda)$ need not always be positive for complex eigenvalues. This may be seen with the help of Figure 5.10, which shows that in two cases the complex eigenvalues move to the origin as $\eta$ approaches infinity. Hence, in this limiting case, $P(\lambda) = P(O) \propto (u^2 - a^2)$. But since the latter factor can be of either sign, it is possible for $\bar{\tau}^*$ and $\bar{T}^*$ together to have signs compared with $\bar{B}_y^*$ which are the same as or opposite those in Table 5.1. Fortunately, this analysis does locate the plane of the spiral sufficiently well so that the existence proof can be completed.

## 7. The Direction of the Integral Curves in the Large

In this section, we develop methods for analyzing the behavior of the integral curves in configuration space away from the singular points. The first problem considered is the direction in which the integral curves cross the null surfaces — if we think of these curves as being developed as x increases. Then we display the regions of configuration space defined by the null surfaces and tabulate the signs of various useful quantities in them. From this, we show how to find the signs of the slopes of the integral curves in each of the regions.

First, consider the F-functions on the right sides of Equations 5.60 through 5.64. The definitions of these functions can be found by comparing these equations with Equations 5.13 through 5.17. Taking the total derivatives of the F's with respect to x,

$$\frac{dF_\tau}{dx} = \left(G^2 - \frac{RT}{\tau^2}\right)\frac{F_\tau}{m_1 G} + \frac{RG}{\tau\kappa} F_T + B_y J_z \tag{5.79}$$

$$\frac{dF_v}{dx} = \frac{G}{m_2} F_v - B_x J_z \tag{5.80}$$

$$\frac{dF_T}{dx} = \frac{RG}{(\gamma - 1)\kappa} F_T + \frac{RT}{\tau}\frac{F_\tau}{m_1 G} - \frac{F_\tau^{\,2}}{m_1 G} - \frac{F_v^{\,2}}{m_2 G} - \frac{J_z}{G}(F_J)_{J_z = 0} \tag{5.81}$$

$$\frac{dF_B}{dx} = \frac{dJ_z}{dx} = \frac{F_J}{\nu G \tau^2} \tag{5.82}$$

$$\frac{dF_J}{dx} = G\tau\mu J_z + \frac{B_y}{m_1} F_\tau - \frac{B_x}{m_2} F_v - \left(\frac{1}{\sigma} + \frac{\nu\tau}{m_1} F_\tau\right)\frac{F_J}{\nu G \tau^2}$$

$$- J_z \frac{d}{dx}\left(\frac{1}{\sigma} + \frac{\nu\tau}{m_1} F_\tau\right) \tag{5.83}$$

The directions in which the integral curves cross each of the null surfaces are then found from

$$\left.\frac{dF_\tau}{dx}\right]_{F_\tau=0} = \frac{RG}{\tau\kappa} F_T + B_y J_z \tag{5.84}$$

$$\left.\frac{dF_v}{dx}\right]_{F_v=0} = - B_x J_z \tag{5.85}$$

$$\left.\frac{dF_T}{dx}\right]_{F_T=0} = \frac{RT}{\tau m_1 G} F_\tau - \frac{F_\tau^{\,2}}{m_1 G} - \frac{F_v^{\,2}}{m_2 G} - \frac{J_z}{G} (F_J)_{J_z=0} \tag{5.86}$$

$$\left.\frac{dJ_z}{dx}\right]_{J_z=0} = \frac{F_J}{\nu G \tau^2} \tag{5.87}$$

$$\left.\frac{dF_J}{dx}\right]_{F_J=0} = G\tau\mu J_z + \frac{B_y}{m_1} F_\tau - \frac{B_x}{m_2} F_v - J_z \frac{d}{dx}\left(\frac{1}{\sigma} + \frac{\nu\tau}{m_1} F_\tau\right) \tag{5.88}$$

The usefulness of these formulas depends upon knowledge of the signs of the F's in each of the regions defined by the null surfaces. The qualitative nature of these regions can be seen sufficiently well by studying the projections on the three-space $\tau$-T-$B_y$ of the null surfaces $F_\tau = 0$, $F_J = 0$, $F_T = 0$. These projections, discussed in Section 3, are illustrated respectively in Figures 2.1, 2.2, and 5.4. The surface $F_J = 0$ exhibits a complicated behavior as $J_z$ moves away from zero. An understanding of this behavior can be obtained by observing, from Equation 5.17 and 5.13, that on the surface $F_\tau = 0$, the pair of hyperbolas of Figure 2.2 move farther and farther apart as $J_z$ increases.

The curve of intersection of the surfaces of Figures 2.1 and 5.4 is shown projected on the $B_y$-$\tau$ plane in Figure 5.5. Then, if all four singular points exist, the intersection of this curve with the hyperbolic cylinder of Figure 2.2 has a projection in the $B_y$-$\tau$ plane similar to that shown in Figure 5.12 and in case 1 of Figure 5.5. It will be sufficient to carry the analysis through in detail using case 1 to differentiate the regions, since the extension of the results on existence and stability to the other two cases will be straightforward.

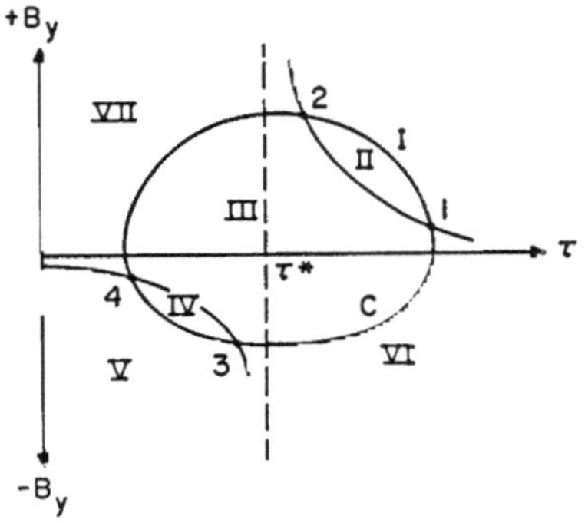


Figure 5.12. Regions defined by the null surfaces

It is clear that these curves divide the $B_y$-τ plane into seven regions, each of which is labeled by a Roman numeral. Considering the temperature axis to protrude upward perpendicular to the paper, one can see from Figures 2.1, 2.2, and 5.4 that any point starting at an arbitrary location in the $B_y$-τ plane at T = 0 and moving upwards parallel to the T-axis must pass through two null surfaces, and hence three regions in T-τ-$B_y$ space. In each of the two-dimensional regions defined by a Roman numeral, these three subregions are denoted by the subscripts 1, 2, and 3 in ascending order. Hence, the three-space τ-T-$B_y$ is separated into 21 regions by the three indicated null surfaces and by the cylinder whose base in the $B_y$-τ plane is the closed curve C of Figure 5.12. Addition of the latter cylinder as a dividing surface is a matter of convenience in classification only. With its removal, it can be seen that the τ-T-$B_y$ space is separated by the null surfaces into only 13 regions, because the following sets of subregions combine: $I_1$, $II_1$; $I_3$, $II_3$; $III_1$, $VI_1$, $VII_1$; $III_3$, $VI_3$, $VII_3$; $IV_1$, $V_1$; $IV_3$, $V_3$.

For each of the above 13 three-dimensional regions, there are two regions in the four-space of T-τ-$B_y$-v, one for $F_v > 0$ and one for $F_v < 0$. Then for each of these 26 four-dimensional regions, there are two regions in the five-space of T-τ-$B_y$-v-$J_z$, one for $J_z > 0$ and one for $J_z < 0$. Thus there are 52 regions in five-space of concern in the analysis of shock existence. The characteristic common to these 52 regions of five-space is that each of them has at least one singular point as a boundary point. Other regions of five-space are not of primary interest.

A great deal of information about the behavior of the integral curves can be obtained if one knows the signs of the F's in the various regions; therefore, the signs of pertinent F's in the three-space regions are given in Table 5.2.

An important factor in the analysis of the behavior of the integral curves is that in a particular region, the projections of the integral curves onto any two-space has a unique sign. For example, division of Equation 5.62 by Equation 5.60 gives the result

$$\frac{dT}{d\tau} = \frac{m_1 G^2}{\kappa} \frac{F_T}{F_\tau} \qquad (5.89)$$

Table 5.2. Signs of the Null-Surface Functions in the Various Regions of Configuration Space

| Region | $F_\tau$ | $F_T$ | $F_J$ | $F_T/F_\tau$ |
|---|---|---|---|---|
| $I_1$ | - | - | + | + |
| $I_2$ | + | - | + | - |
| $I_3$ | + | + | + | + |
| $II_1$ | - | - | + | + |
| $II_2$ | - | + | + | - |
| $II_3$ | + | + | + | + |
| $III_1$ | - | - | - | + |
| $III_2$ | - | + | - | - |
| $III_3$ | + | + | - | + |
| $IV_1$ | - | - | + | + |
| $IV_2$ | - | + | + | - |
| $IV_3$ | + | + | + | + |
| $V_1$ | - | - | + | + |
| $V_2$ | + | - | + | - |
| $V_3$ | + | + | + | + |
| $VI_1$ | - | - | - | + |
| $VI_2$ | + | - | - | - |
| $VI_3$ | + | + | - | + |
| $VII_1$ | - | - | - | + |
| $VII_2$ | + | - | - | - |
| $VII_3$ | + | + | - | + |

This specific slope is particularly useful because its sign is fixed in each region of the three-space $T$-$\tau$-$B_y$; hence, its sign is given in Table 5.2. For example, in Regions $II_2$, $III_2$, and $IV_2$, $dT/d\tau < 0$, indicating that temperature and density increase together. Since this is to be expected in real shocks, one strongly suspects that regions like these may contain shock-layer curves, but that regions for which $dT/d\tau > 0$ will not. It will also be of importance to recognize from this discussion that within each region the projections of the integral curves on any two-space are single-valued functions.

## 8. Fast Shocks

The stage has now been prepared for the primary task of the latter half of this monograph — to determine which of the singular points are connected by integral curves, and to determine if there can be more than one such curve connecting any pair of singular points. In other words, the question is whether unique shock-layer curves exist. In the present section, the discussion will be limited to transitions between $SP_1$ and $SP_2$, that is, to fast-shock transitions. The addition of current inertia makes this discussion quite difficult and prolonged in comparison with the proof of existence of a unique fast shock given by Germain.[13] By what seems to be a mathematical fortuity, Germain was able to carry his proof through to completion using much less than the total amount of information available. Mother Nature, however, is not always so kind; current inertia is present and removes the niceties of the problem in such a way that all possible information must be brought to bear on the problem if the behavior of the integral curves is to be understood.

This discussion is opened most logically by considering the $\lambda$'s at $SP_1$ and $SP_2$. Figure 5.11 shows that at $SP_1$, four of the $\lambda$'s are positive and one ($\lambda_1$) is negative, and that they are all always real. Thus, $SP_1$ can be either an upstream or a downstream singular point, although from Section 3 of Chapter 2 we know that the second law of thermodynamics can be satisfied only if $SP_1$ turns out to be the upstream point, and $SP_2$ the downstream point. From Figures 5.1 and 5.2, it can be seen that $SP_1$ is a saddle point in any plane containing $\vec{\lambda}_1$ and an upstream node in any plane perpendicular to $\vec{\lambda}_1$. Note also, from Equation 5.7 and Figure 5.1, that since $\lambda_2$ is the smallest of the positive eigenvalues, in any plane orthogonal to $\vec{\lambda}_1$ and containing $\vec{\lambda}_2$, there will be an infinite number of integral curves coming from $SP_1$ in the direction of $\vec{\lambda}_2$.

Consider $SP_2$. Figure 5.11 shows that three $\lambda$'s are always positive and real and that two are either real and negative or complex with negative real parts. Referring to Figures 5.1, 5.2, and 5.3, we can see that in the plane of $\vec{\lambda}_1$ and $\vec{\lambda}_2$, $SP_2$ is either a downstream node or a downstream spiral point. In any plane perpendicular to both of these eigenvectors, it is an upstream node; hence, the shock-layer curve can exist only as an integral curve approaching $SP_2$ in the plane of $\vec{\lambda}_1$ and $\vec{\lambda}_2$. Also, from Equation 5.7 and Figure 5.1, a single integral curve approaches the singular point in the direction of the $\vec{\lambda}$ for which $\lambda$ has the largest magnitude, hence in the direction of $\vec{\lambda}_1$ if $\lambda_1$ is real. An infinite number of integral curves approach in the direction of $\vec{\lambda}_2$.

Now consider the $\vec{\lambda}$'s at $SP_1$. The relative signs of the components of $\vec{\lambda}_1$, $\vec{\lambda}_2$, and $\vec{\lambda}_5$ are given in Table 5.1. From Table 5.1 and the discussion below Equation 5.65, it can easily be deduced that these eigenvectors lie in the following five-dimensional regions:

$$\vec{\lambda}_1: F_T > 0,\ F_\tau < 0,\ F_V > 0,\ J_z > 0,\ F_J < 0 \tag{5.90}$$

$$\vec{\lambda}_2: F_T > 0,\ F_\tau < 0,\ F_V > 0,\ J_z > 0,\ F_J > 0 \tag{5.91}$$

$$\vec{\lambda}_5: F_T > 0,\ F_\tau > 0,\ F_V < 0,\ J_z > 0,\ F_J > 0 \tag{5.92}$$

Then, referring to Table 5.2 and Figure 5.12, we can see that the projection of $\vec{\lambda}_1$ into the three-space $T$-$B_y$-$\tau$ lies in $III_2$ if the signs of the components of $\vec{\lambda}_1$ are as shown in Table 5.1, and in $I_2$ if they are opposite. Similarly, one branch of $\vec{\lambda}_2$ lies in Region $II_2$ and one in $VI_2$, and one branch of $\vec{\lambda}_5$ lies in $I_3$-$II_3$ and one in $III_1$-$VI_1$. Note also that the two colinear $\vec{\lambda}_1$'s point toward $SP_1$, whereas the $\vec{\lambda}_2$'s and $\vec{\lambda}_5$'s point away from $SP_1$.

Next, let us examine the $\vec{\lambda}$'s at $SP_2$. From Table 5.1 the regions of five-space in which $\vec{\lambda}_1$, $\vec{\lambda}_2$, and $\vec{\lambda}_5$ lie are as follows:

$$\vec{\lambda}_1, \vec{\lambda}_2: \quad F_T > 0,\ F_\tau < 0,\ F_V > 0,\ J_z > 0,\ F_J < 0 \tag{5.93}$$

$$\vec{\lambda}_5: \quad F_T > 0,\ F_\tau > 0,\ F_V < 0,\ J_z > 0,\ F_J > 0 \tag{5.94}$$

Again, the peculiar orthogonality properties of the five-space of this problem permit one single pair of $\vec{\lambda}$'s to lie in a single five-space region. From Table 5.2 and Figure 5.12, the projections of $\vec{\lambda}_1$ and $\vec{\lambda}_2$ into the three-space $T$-$B_y$-$\tau$ lie in $III_2$ if the signs of their components are given in Table 5.1, and in $I_2$ if they are opposite. Similarly, $\vec{\lambda}_5$ lies in $I_3$-$II_3$ or in $III_1$-$VII_1$.

The next step is to consider the slopes of the integral curves in the $T$-$B_y$-$\tau$ space. Although this does not give the full picture of the integral curves, it is important because the existence of an integral curve going from $SP_1$ to $SP_2$ in three-space must, by continuity, be a prerequisite to the existence of integral curves going from $SP_1$ to $SP_2$ in five-space. It is the projection of these five-space curves that is observed in three-space. These remarks hold true in this problem because the actual five-space singular points lie in the $T$-$B_y$-$\tau$ three-space considered. The most useful slope to study is $dT/d\tau \propto F_T/F_\tau$, because its sign, given in

Table 5.2, is fixed in each of the 13 regions of $T$-$B_y$-$\tau$ space. Since real shocks are compressive, and the gas temperature rises through them, it is clear that $dT/d\tau$ must be at least predominantly — if not always — negative in the shock layer. Table 5.2 shows that $dT/d\tau$ is negative only in the regions with the subscript 2, that is, in the regions between the $F_T$ and $F_\tau$ null surfaces. Now, if we look at Figure 5.12 from a purely mathematical standpoint, we can see that integral curves from $SP_1$ coming from any $\vec{\lambda}$'s which may exist in $I_3$, $II_3$, $III_3$, or $VI_3$ must move upward and to the right. Those which do not contact the null surface $F_T = 0$ continue on to infinity and cannot connect with other singular points. Some of these integral curves will, however, cross $F_T = 0$. Their slopes then become negative, and they move downward but still to the right and away from the other singular points. Some of these curves may eventually contact the null surface $F_\tau = 0$. The sign of $dT/d_\tau$ will change again, and they will move downward and to the left until they contact the plane $T = 0$. Thus, it is clear that no integral curves starting from $SP_1$ in the subscript-3 regions can correspond to shocks. Similarly, it can be seen that integral curves starting from $SP_1$ in the subscript-1 region cannot correspond to shocks.

By arguments like those given above, it is evident that integral curves starting from $SP_1$ in the Regions $I_2$ and $VI_2$ also cannot correspond to shocks, because these integral curves start off in the wrong direction and can never get back to the singular points. (The reason $I_2$ is completely excluded will become clearer later in the proof.) This leaves Regions $II_2$ and $III_3$ as the only ones from which shock-layer curves can begin. These considerations also show that the integral curve corresponding to a shock-layer curve (if there is one) must stay in $II_2$-$III_2$, for if it hits either the upper or lower boundary and if it can pass through, it acquires a positive slope $dT/d_\tau$ and must go off either to $T = \infty$ or $T = 0$.

By analyzing the slopes of integral curves in regions around $SP_2$, the same conclusion is reached: Integral curves corresponding to shocks can arrive at $SP_2$ only via Regions $II_2$ and $III_2$. But only in $III_2$ do the $\vec{\lambda}$'s point toward $SP_2$; therefore, integral curves corresponding to shocks can arrive only in $III_2$..

Now consider the orthogonality properties of the five $\vec{\lambda}_i$. It was shown in Section 4 that in the five-space $T$-$\tau$-$B_y$-$J_z$-$v$, $\vec{\lambda}_2$, $\vec{\lambda}_3$, $\vec{\lambda}_4$, and $\vec{\lambda}_5$ are orthogonal in the ordinary sense, and $\vec{\lambda}_1$ is skewed with respect to these four vectors. But the set of equations like 5.65 shows that near a singular point, the five surfaces $F_i = 0$ form a set of five orthogonal planes; i.e., the $dF_i$ have the same orthogonality properties as the eigenvector components on the left side. Just as three orthogonal planes divide three-space into octants, so these five four-dimensional "planes" divide five-space into 32 regions. No two of the vectors $\vec{\lambda}_2 \cdots \vec{\lambda}_5$ can lie in any one of these regions; hence, in particular, neither $\vec{\lambda}_3$ nor $\vec{\lambda}_4$ at $SP_1$ can lie in the region defined by Inequalities 5.91, nor can they, at $SP_2$, lie in the region defined by Inequalities 5.93. If either $\vec{\lambda}_3$ or $\vec{\lambda}_4$ is in $II_2$ at $SP_1$, it must lie in $F_v < 0$. This can be explained by the following reasoning: Since $F_J > 0$ in $II_2$, $dJ_z/dx > 0$ there, by Equation 5.64, and hence points moving in the direction of positive eigenvectors must move into $J_z > 0$. Then, the only way for $\vec{\lambda}_3$ or $\vec{\lambda}_4$ to differ from $\vec{\lambda}_2$, which they must do by orthogonality, would be for $F_v$ to be negative.

We shall show later that it will be necessary for $F_v$ to be of the same sign at both ends of the transition; hence, the shock-layer curve — if it exists — must leave $SP_1$ via $\vec{\lambda}_2$ and terminate at $SP_2$ via either $\vec{\lambda}_1$ or $\vec{\lambda}_2$.

Next, let us examine the slopes $dB_y/d\tau$ and $dB_y/dT$ in $T$-$\tau$-$B_y$ space. By means of Equations 5.60, 5.62, and 5.63, the signs of these slopes are found from

$$
\left.
\begin{aligned}
\frac{dB_y}{d\tau} &\propto \frac{J_z}{F_\tau} \\[2em]
\frac{dB_y}{dT} &\propto \frac{J_z}{F_T}
\end{aligned}
\right\}
\tag{5.95}
$$

If $J_z > 0$ in $II_2$ and $III_2$, $dB_y/d\tau < 0$ and $dB_y/dT > 0$. Both of these slopes are correct if the integral curves move monotonically from $SP_1$ to $SP_2$. Also Equations 5.91 and 5.93 show that $J_z > 0$ near singular points. It is necessary, then, to determine whether it is possible for $J_z$ to become negative somewhere along an integral curve which remains in the combined Region $II_2$-$III_2$. The key to this analysis is the fact that on $F_J = 0$ (the border between $II_2$ and $III_2$), $dJ_z/dx = 0$.

Assume that a projected integral curve can exist in $II_2$-$III_2$ such that $J_z(x)$ is of the form shown in Figure 5.13. Let this integral curve start from $SP_1$ along $\vec{\lambda}_2$, that is, in $II_2$. Then, referring to Figure 5.12 and Relations 5.95, we see that it starts out with negative slope in the $B_y$-$\tau$ plane and eventually must cross $F_J = 0$. It then continues into $III_2$, and when $J_z = 0$, its slope $dB_y/d\tau = 0$. As $x$ increases further, $J_z$ becomes negative and $dB_y/d\tau$ positive. The curve must then extend down to and across the other branch of $F_J = 0$. Then, following Figure 5.13, we find that while it is in $IV_2$, $J_z$ vanishes again and with it $dB_y/d\tau$. Continuing in this way, we see that the integral curve continues toward the left and can never return to $SP_2$. Hence, if

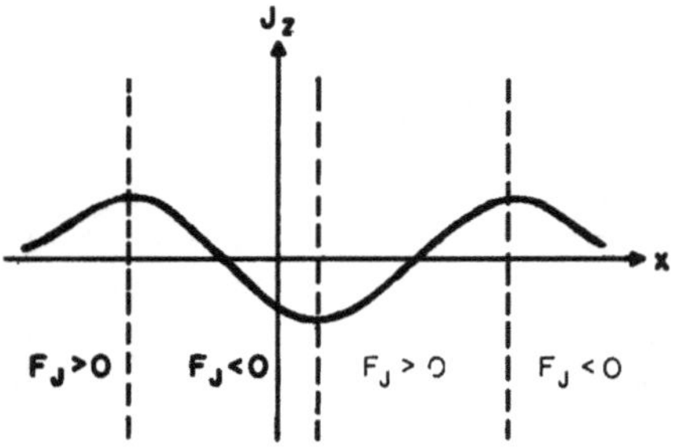


Figure 5.13 Postulated form of $J_z(x)$

there are integral curves connecting $SP_1$ and $SP_2$, $J_z > 0$ throughout. Thus, we can prescribe integral curves in five-space with a function $J_z(x)$, as in Figure 5.13, but they cannot correspond to fast-shock layers.

One can prescribe integral curves in five-space for which $J_z = 0$, and these curves can correspond to fast shocks. Hence, let us consider integral curves which remain in the five-space region $R_5$, defined, from 5.91 and 5.93, by

$$F_T > 0,\ J_z > 0,\ F_v > 0,\ F_\tau < 0 \qquad (5.96)$$

It is clear from Equation 5.64 that all possible shock-layer curves must cross $F_J = 0$ an odd number of times. It is also evident from 5.95 that the projection of these curves onto the $B_y$-$\tau$ plane must remain in the region where $\tau > \tau_2$ and $B_y > B_{y_1}$. (The subscripts refer to the singular points.)

Now consider the behavior of the shock-layer curves as they cross the boundaries of the region defined by 5.96. This behavior is determined by Equations 5.84 through 5.87. From Equations 5.84 and 5.85,

$$\left.\frac{dF_\tau}{dx}\right|_{F_\tau=0} > 0, \qquad \left.\frac{dF_v}{dx}\right|_{F_v=0} < 0 \qquad (5.97)$$

at the boundary of $R_5$. This says that all integral curves on the boundaries $F_\tau = 0$ and $F_v = 0$ of $R_5$ leave $R_5$. The behavior at the other two boundaries is more complicated. Equation 5.86 shows that

$$\left.\frac{dF_T}{dx}\right|_{F_T=0} < 0 \qquad (5.98)$$

if $(F_J)_{J_z=0} > 0$. But, by comparing Equations 5.17 and 5.64, we obtain

$$(F_J)_{J_z=0} = G\tau B_y - vB_x \qquad (5.99)$$

This surface is projected onto the plane $F_v = 0$ by substituting for v from Equation 5.19, the result of which is the surface of Equation 5.22 (Figure 2.2). But this is the hyperbolic cylinder shown in cross section in Figure 5.12. Hence, $(F_J)_{J_z=0}=0$ separates the projected regions $II_2$ and $III_3$, and it is clear that within projected $II_2$, Inequality 5.98 holds, and the integral curves can only leave $R_5$ through $F_T=0$. In projected $III_3$, $(F_J)_{J_z=0}<0$; hence, Equation 5.86 shows that at any point there is a $J_z$ for which the direction of the inequality in 5.98 is reversed. In this qualitative analysis, it does not seem possible to determine whether the inequality can actually reverse itself; therefore, that possibility must be assumed. This means that in projected $III_2$, there will be regions in $R_5$ where integral curves can only leave $R_5$ through the surface $F_T = 0$, and there also may be regions in which integral curves enter $R_5$ through $F_T = 0$. Those curves which do leave through $F_T = 0$ cannot correspond to shocks, because they thereupon acquire a positive slope $dT/d\tau$ and move off in a direction in which they cannot reach $SP_2$. Curves which enter $III_2$ through $F_T = 0$ cannot correspond to shocks either, because they must have come up through the plane $T = 0$; however, there is still room in $III_2$ for integral curves connecting points $SP_1$ and $SP_2$.

Finally, the behavior of integral curves passing through $J_z = 0$ must be discussed. Equation 5.87 shows that these curves enter $R_5$ when $F_J > 0$ and leave $R_5$ when $F_J < 0$; that is, they enter when $dJ_z/dx > 0$ and leave when $dJ_z/dx < 0$. It has already been shown, however, that integral curves which actually cross $J_z = 0$ cannot correspond to transition between $SP_1$ and $SP_2$.

We can obtain one more useful fact before attempting to trace transitions from $SP_1$ to $SP_2$. Since, at this point, shock transitions apparently can arrive at $SP_2$ in the direction of either $\vec{\lambda}_1$ or $\vec{\lambda}_2$, when they are real, it is worth while to know the relative orientation of these two vectors in $III_2$. This information can be found from Expressions 5.67 through 5.70 and by referring back to Equation 5.25. From Expression 5.67,

$$\frac{\overline{\tau}^*}{\overline{T}^*} \propto \lambda - \lambda_\kappa$$

At $SP_2$, $\lambda_1$ and $\lambda_2$ are both negative and $|\lambda_1| > |\lambda_2|$. Hence,

$$\left|\frac{\overline{\tau}^*}{\overline{T}^*}\right|_1 > \left|\frac{\overline{\tau}^*}{\overline{T}^*}\right|_2 \tag{5.100}$$

From Expression 5.68, $\overline{B}_y^*\lambda \propto \overline{J_z}^*$. Therefore, at $SP_2$,

$$\left|\frac{\bar{J}_z^*}{\bar{B}_y^*}\right|_1 > \left|\frac{\bar{J}_z^*}{\bar{B}_y^*}\right|_2 \tag{5.101}$$

From Expression 5.70,

$$\frac{\bar{T}^*}{\bar{B}_y^*} \propto \frac{1}{P(\lambda)}$$

But $P(\lambda_1) > P(\lambda_2)$; hence,

$$\left|\frac{\bar{T}^*}{\bar{B}_y^*}\right|_1 < \left|\frac{\bar{T}^*}{\bar{B}_y^*}\right|_2 \tag{5.102}$$

From Expression 5.69,

$$\frac{\bar{T}^*}{\bar{B}_y^*} \propto \frac{\lambda - \lambda_\kappa}{P(\lambda)}$$

But, from Equation 5.73, $P(\lambda)$ is of the form

$$P(\lambda) = c_1 \left[(\lambda_{m_1} - \lambda)(\lambda_\kappa - \lambda) - c_2\right]$$

where $c_1$ and $c_2$ are positive constants. Using this expression, we can easily prove that

$$\left|\frac{\lambda_1 - \lambda_\kappa}{P(\lambda_1)}\right| < \left|\frac{\lambda_2 - \lambda_\kappa}{P(\lambda_2)}\right|$$

Hence,

$$\left|\frac{\bar{T}^*}{\bar{B}_y^*}\right|_1 < \left|\frac{\bar{T}^*}{\bar{B}_y^*}\right|_2 \tag{5.103}$$

From the known signs of the components of $\vec{\lambda}_1$ and $\vec{\lambda}_2$ at $SP_2$, and from Inequalities 5.100 through 5.103, we can see that $\vec{\lambda}_1$ and $\vec{\lambda}_2$ lie relative to each other in $III_2$, as shown in the three projected views in Figure 5.14. For reference, the relative location of $\vec{\lambda}_2$ at $SP_1$ is also shown.

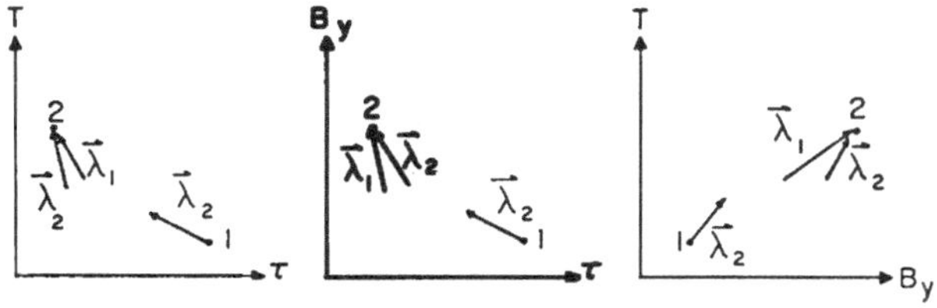

Figure 5.14. Relative orientation of $\vec{\lambda}_1$ and $\vec{\lambda}_2$ at $SP_2$

With the help of all of the above information, it is now possible to trace integral curves in order to determine whether fast shocks exist. It is most convenient to start from $SP_2$ and trace these curves backwards, i.e., in the negative-x direction. Starting backwards along either $\vec{\lambda}_1$ or $\vec{\lambda}_2$ or along a spiral if $\lambda_1$ and $\lambda_2$ are complex conjugates, some of the curves head in the general direction of $SP_1$, since $J_z > 0$. They can never leave $R_5$ through either $F_\tau = 0$ or $F_v = 0$, and while $F_J < 0$, they cannot leave through $J_z = 0$. There is a possibility, however, that the integral curves could leave through $F_T = 0$—though if they do, they cannot reach $SP_1$. But $R_5$ is everywhere dense with integral curves; and, in fact, in the projection of $R_5$ onto the three-space $\tau$-T-$B_y$, integral curves overlap each other, because at each point the slope depends on $J_z$ and v. Therefore, for every integral curve in $III_2$ which leaves through $F_T = 0$, there is one lying at a higher temperature that will move further away from $SP_2$ within $III_2$. Continuing in this way, there are integral curves which reach $(F_J)_{J_z=0} = 0$ before they pass through $F_T = 0$. But after passing through $(F_J)_{J_z=0} = 0$, they can no longer leave $R_5$ through $F_T = 0$. Some of the curves pass through $(F_J)_{J_z=0} = 0$ and then through $F_J = 0$ in that order. After this, they can leave $R_5$ only through $J_z = 0$ because now $F_J > 0$. Those that do leave cannot correspond to shocks. Those that do not leave $R_5$ must terminate at $SP_1$ in the direction of $\vec{\lambda}_2$.

When $\lambda_1$ and $\lambda_2$ are complex at $SP_2$, the above argument follows without modification if the plane of the resulting spiral intercepts the region of the real $\vec{\lambda}_1$ and $\vec{\lambda}_2$. It was shown in Section 6, however, that at least one vector in the plane of the spiral can shift from the region defined by Inequalities 5.93 into the region

$$F_T < 0, \; F_\tau > 0, \; F_v > 0, \; J_z > 0, \; F_J < 0$$

This is not a necessary condition for the plane of the spiral to cease to intercept the region of Inequalities 5.93, but it opens that possibility. If it does happen, the plane of the spiral will move into the projected regions $II_2$ and $VII_2$ of Figure 5.12. But in this case an integral curve from $SP_2$ can be traced back to $SP_1$ in the same manner as before; hence, the existence proof is still valid.

From the preceding argument, it is clear that fast shocks exist, but it is not obvious that they are unique. The proof of uniqueness will now be given. This proof is more difficult than in the case in which current inertia is neglected, because there are not only an infinite number of integral curves leaving $\vec{\lambda}_2$ at $SP_1$, but also an infinite number arriving at $SP_2$ along $\vec{\lambda}_2$. When current inertia is neglected $\vec{\lambda}_1$ disappears, and only one integral curve coming from the direction of $SP_1$ can connect with $\vec{\lambda}_2$ at $SP_2$. The following proof depends on the topological properties of the integral curves in the space of $\vec{\lambda}_1$, $\vec{\lambda}_2$ and either $\vec{\lambda}_3$, $\vec{\lambda}_4$, or $\vec{\lambda}_5$. (It should be recalled that $\lambda_3$, $\lambda_4$, and $\lambda_5$ are all positive at both $SP_1$ and $SP_2$.)

Consider the eigenvector space of $\vec{\lambda}_1$, $\vec{\lambda}_2$, and, say, $\vec{\lambda}_3$ at both $SP_1$ and $SP_2$. As usual, the case of real eigenvectors will be discussed first. In the imagination, these triads are bent and twisted into the form shown in Figure 5.15, where two possible configurations of integral curves connecting the region around $SP_1$ to the region around $SP_2$ are shown. From the linearized study in the neighborhood of the singular points, it is known that $\vec{SP}_1$ is a saddle point in the planes $(\vec{\lambda}_1, \vec{\lambda}_2)$ and $(\vec{\lambda}_1, \vec{\lambda}_3)$ and an upstream node in $(\vec{\lambda}_2, \vec{\lambda}_3)$. It is known that $\vec{\lambda}_3$ is the exceptional direction in the $(\vec{\lambda}_2, \vec{\lambda}_3)$ plane, so that an infinite number of curves leave in the direction of $\vec{\lambda}_2$ in that plane. Moreover, $SP_2$ is a saddle point in the $(\vec{\lambda}_1, \vec{\lambda}_3)$ and $(\vec{\lambda}_2, \vec{\lambda}_3)$ planes, and a downstream node in the $(\vec{\lambda}_1, \vec{\lambda}_2)$ plane, in which $\vec{\lambda}_1$ is the exceptional direction. With this information, and with the two triads oriented as shown in Figure 5.15a, one can see that the integral curves fall naturally into the indicated pattern. The solid integral curves are in the vertical plane and the dotted curves in the horizontal plane. Although the linearized solution breaks down

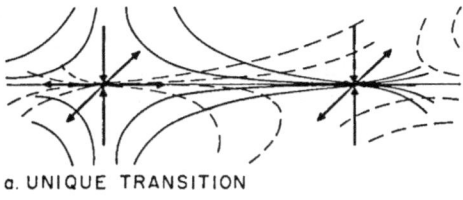

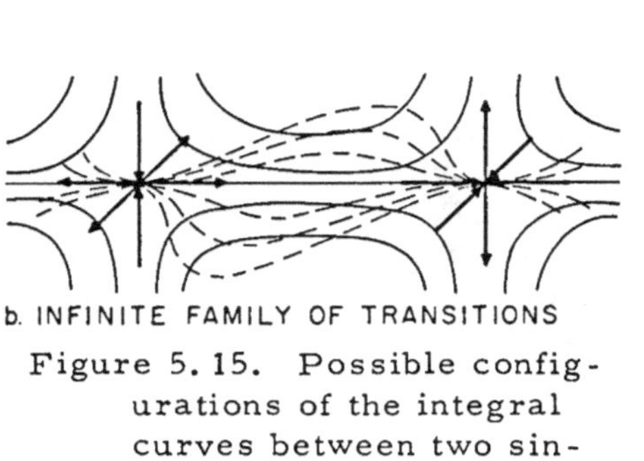


Figure 5.15. Possible configurations of the integral curves between two singular points

in the region away from the singular points, the indicated smooth behavior of the integral curves is justified by their monotonic behavior in each of the regions formed by the null surfaces. In the $T$-$\tau$-$B_y$-$J_z$-$v$ configuration space, the two triads have a different relative orientation, but the integral curves can be thought of as being bent and twisted along with them. Thus, the requirement that $dJ_z/dx = 0$ on $F_J = 0$, across which these integral curves must go, can easily be met with the topological configuration of Figure 5.15a. If this is the actual configuration, it is clear that there is a unique transition.

If the triad at $SP_2$ is rotated 90° about $\vec{\lambda}_2$, the question must be answered as to whether the family of integral curves of Figure 5.15a merely twists, or whether the actual configuration is as shown in Figure 5.15b. If the latter configuration prevails, clearly an infinite number of transitions is indicated, and there is no unique shock-layer curve. The characteristic of Configuration b which will be used to show that it is not possible is that in the horizontal plane shown, each successive member of the family of integral curves connecting $SP_1$ to $SP_2$ must bulge out farther away from the $\vec{\lambda}_2$-axis. From Figure 5.12 it can be seen that eventually these curves will have to penetrate the $F_T$ and/or $F_\tau$ null surfaces. But then they leave Region $II_2$-$III_2$, and it has already been shown that if they do, they can never come back. For this reason the configuration of Figure 5.15b is not possible. The topological configuration of Figure 5.15a is therefore the only possible one; hence, there is only one integral curve connecting $SP_1$ and $SP_2$.

In the case when $\lambda_1$ and $\lambda_2$ at $SP_2$ are complex, $SP_2$ is a spiral point in the plane perpendicular to $\vec{\lambda}_3$. For the orientation of Figure 5.15a, however, there still can be only one curve connecting $SP_1$ to $SP_2$. Curves which leave $SP_1$ in the ($\vec{\lambda}_2$, $\vec{\lambda}_3$) plane and curve toward the reader must acquire greater and greater slope in the direction of $\vec{\lambda}_3$ at $SP_3$; hence, they go off to infinity. For the orientation of Figure 5.15b the curves must bulge out exactly as in the case of real eigenvalues, and for the same reason this configuration cannot be correct.

From the arguments given above, this section can be concluded positively with the statement that a unique fast-shock layer curve exists.

## 9. Slow Shocks

The argument for existence of a unique slow shock will have to make much greater use of the topological properties of integral curves in configuration space remote from the singular points, because the positions of the eigenvectors at $SP_3$, from which integral curves can can come, cannot be determined algebraically from the linear analysis. We shall present these arguments in the sequel, but first we shall discuss known properties of the eigenvalues and eigenvectors at $SP_3$ and $SP_4$ in relation to the null surfaces and determine the regions at $SP_3$ in which eigenvectors must lie to give rise to transition curves.

Consider the eigenvalues and eigenvectors at $SP_3$. Figure 5.11 shows that the only positive $\lambda$'s there are the real quantities $\lambda_3$, $\lambda_4$, $\lambda_5$ ; hence, any transitions from $SP_3$ must begin along either $\vec{\lambda}_3$, $\vec{\lambda}_4$, $\vec{\lambda}_5$. From Table 5.1, it can be seen that when they are real, $\vec{\lambda}_1$ and $\vec{\lambda}_2$ lie in the region

$$F_T > 0,\ F_\tau < 0,\ F_v < 0,\ J_z < 0,\ F_J > 0 \qquad (5.104)$$

and that $\vec{\lambda}_5$ lies in

$$F_T > 0,\ F_\tau > 0,\ F_v > 0,\ J_z < 0,\ F_J < 0 \qquad (5.105)$$

Thus, in the subspace of Figure 5.12, $\vec{\lambda}_1$ and $\vec{\lambda}_2$ lie in $IV_2$ and $VI_2$, and $\vec{\lambda}_5$ lies in the combined regions $III_3$-$VI_3$ and $IV_1$-$V_1$. In the case of $\vec{\lambda}_1$ and $\vec{\lambda}_2$, note from Table 5.1 that both $\bar{\tau}^*$ and $\bar{B}^*_y$ are negative. Since $F_J > 0$, it does not appear from Figure 5.12 that this situation is possible until one realizes that this figure is a projection from five-space, and that if it were projected onto planes of constant $J_z$ less than zero, the hyperbolas would be closer together. All that we know about $\vec{\lambda}_3$ and $\vec{\lambda}_4$ is that they cannot lie in the regions which defy the orthogonality requirements, those being the regions described by Inequalities 5.104 and 5.105 with the signs of $J_z$ and $F_J$ reversed. In this analysis, we must therefore assume at this point that they can lie in any of the three-space regions around $SP_3$ in Figure 5.12.

At $SP_4$, Figure 5.11 shows that $\lambda_1$ and $\lambda_2$ are negative and real or complex, $\lambda_3$ is negative and real, and $\lambda_4$ and $\lambda_5$ are positive and real. The nature of the integral curves near $SP_4$ in the planes formed by the $\vec{\lambda}$'s can be deduced with the help of Figures 5.1, 5.2, and 5.3. Any transitions to $SP_4$ must clearly end along $\vec{\lambda}_1$, $\vec{\lambda}_2$, or $\vec{\lambda}_3$. When $\vec{\lambda}_1$ and $\vec{\lambda}_2$ are real, Figure 5.1 shows that $\vec{\lambda}_1$ is the only eigenvector along which a single integral curve terminates; $\vec{\lambda}_2$ and $\vec{\lambda}_3$ accept a single and double infinity of integral curves, respectively. When $\lambda_1$ and $\lambda_2$ are complex, a single infinity of integral curves spiral into $SP_4$ in the plane determined by $\mathrm{Re}(\vec{\lambda}_1)$ and $\mathrm{Im}(\vec{\lambda}_1)$.

The eigenvectors $\vec{\lambda}_1$, $\vec{\lambda}_2$, and $\vec{\lambda}_5$ at $SP_4$ lie in the five-space regions defined by the Inequalities 5.104 and 5.105 and, from Table 5.1, $\vec{\lambda}_3$ lies in the region

$$F_T > 0,\ F_\tau < 0,\ F_v > 0,\ J_z > 0,\ F_J < 0 \qquad (5.106)$$

Under the assumption that a transition between $SP_3$ and $SP_4$ exists, we shall now show that it must come from an eigenvector in the Region $IV_2$ at $SP_3$, say $\vec{\lambda}_a$. First, by an argument based on Equation 5.95 and used in Section 8, it is clear that $\vec{\lambda}_a$ must lie in either $III_2$, $IV_2$, or $V_2$. Suppose that it lies in $III_2$. Then, since $F_J < 0$ there, the curve moves into the region $J_z < 0$. At $SP_4$, $\vec{\lambda}_3$ lies in $F_J < 0$, $J_z > 0$; hence, to terminate at this eigenvector, $J_z$ must pass through zero an odd number of times. In Figure 5.12, this means that $dB_y$ must vanish an odd number of times along the transition curve. But then this curve cannot terminate at $\vec{\lambda}_3$, because to do so, $dB_y$ must vanish along it an even number of times. To terminate at $\vec{\lambda}_1$ or $\vec{\lambda}_2$, the transition curve would have to pass through a point $dB_y = 0$ an odd number of times, but $J_z < 0$ at both end points, which would require $dB_y$ to vanish an even number of times. From this argument, it is clear that $\vec{\lambda}_a$ cannot lie in $III_2$. The fact that $\vec{\lambda}_a$ cannot lie in $V_2$ may be seen directly from the first of Equations 5.95. In $V_2$, $F_\tau > 0$; thus, if the slope of $\vec{\lambda}_a$ in the $B_y$ - $\tau$ plane is positive, $J_z > 0$, and consequently $dB_y > 0$ when $d\tau > 0$. This means that the integral curve would have to move, in Figure 5.12, to the right of $SP_3$ and away from $SP_4$. Similarly, if the slope of $\vec{\lambda}_a$ in the $B_y$ - $\tau$ plane is negative, $J_z < 0$, and consequently $dB_y < 0$ when $d\tau > 0$, in which case the curve moves downward and to the right, away from $SP_4$. In neither of these cases can the integral curve turn around in such a way as to terminate at $SP_4$. Only for $\vec{\lambda}_a$ in $IV_2$ can all requirements of a transition between $SP_3$ and $SP_4$ be met; hence, if this transition exits, $\vec{\lambda}_a$ at $SP_3$ must lie in $IV_2$.

If the transition from $SP_3$ to $SP_4$ (assuming one exists) goes to $\vec{\lambda}_3$ at $SP_4$, $J_z > 0$ throughout, by an argument identical to one given in Section 8 and based on observation of the behavior of $J_z(x)$. Then, Equation 5.85 shows that integral curves in the region $J_z > 0$ can only leave the region $F_v > 0$; hence, an integral curve connecting $SP_3$ and $SP_4$ and terminating along $\vec{\lambda}_3$ must lie wholly in $F_v > 0$. With this information, it is clear that if $\vec{\lambda}_a$ does initiate a shock which terminates along $\vec{\lambda}_3$, it must lie in the five-space region

$$F_T > 0, \quad F_\tau < 0, \quad F_v > 0, \quad J_z > 0, \quad F_J > 0 \qquad (5.107)$$

Thus, from Inequality 5.66, $\lambda_a < \lambda_{m_2}$; from 5.67, $\lambda_a < \lambda_\kappa$; and from 5.70, $P(\lambda_a) < 0$ — all of which can easily be satisfied by either $\lambda_3$ or $\lambda_4$. If the transition terminates along either $\vec{\lambda}_1$ or $\vec{\lambda}_2$, $J_z$ passes through zero at least once; hence, the sign of $dF_v/dx$ from Equation 5.85 is indefinite. Consequently, the sign of $F_v$ in the region in which $\vec{\lambda}_a$ is located is not definitely determined. In the other dimensions, however, $\vec{\lambda}_a$ still lies in the region indicated by Inequalities 5.107.

We have now shown that a transition from $SP_3$ to $SP_4$ is possible if there is an eigenvector $\vec{\lambda}_a$ at $SP_3$ in the region defined by Inequalities 5.107 (or possibly with the sign of $F_v$ reversed); however, even if $\vec{\lambda}_a$ does exist, the proof of existence of the shock layer, along the lines used for fast shocks, cannot be carried through in this case because the sign of $dF_T/dx$, as determined by Equation 5.84, is not definitely known. Moreover, it has been shown that if the shock layer exists, it could terminate along either $\vec{\lambda}_3$ or one of the pair $\vec{\lambda}_1$, $\vec{\lambda}_2$. In the former case, there can be no downstream spatial oscillations because $\lambda_3$ is always real, and also $J_z > 0$ throughout the transition. In the latter case, there can be spatial oscillations, and $J_z$ must change sign somewhere in the transition. To determine the actual behavior, it is necessary to look more deeply into the topological properties of the integral curves, and we shall demonstrate that the properties of the linearized solution near the singular points determine, to a much greater degree than at first appears possible, the total topological behavior of the integral curves. We shall explain the topological properties in terms of four concepts.

The first concept can be stated as a theorem:

> Points on adjacent integral curves moving in the direction of increasing x must move in the same direction in configuration space; i.e., one can think of the integral curves as stream lines in a potential flow and the singular points as sources and sinks.

The proof of this theorem follows directly from the differential equations of the shock layer, Equations 5.13 through 5.17. At a given point in configuration space the right sides are fixed; thus, if dx is assigned a certain magnitude and sign, the differentials in the numerator define a vector pointing in a certain definite direction. If the sign of dx is reversed, the vector reverses direction. Hence, an increase in dx always corresponds to motion in a certain definite direction and provides a definite restriction on the way that integral curves coming from the region around one singular point can mesh with those from another.

A second concept is directly related to the above-mentioned mesh of integral curves and has already been introduced in connection with Figure 5.15. In a plane, there are three basic configurations, shown in Figure 5.16, which result in transitions from one singular point to another. Configurations in which there are only positive eigenvectors at both points, or only negative ones, clearly can produce no transitions. Incidental modifications of these configurations are obtained by changing or reversing the ratio between the λ's at a given point, or by reversing the signs of all

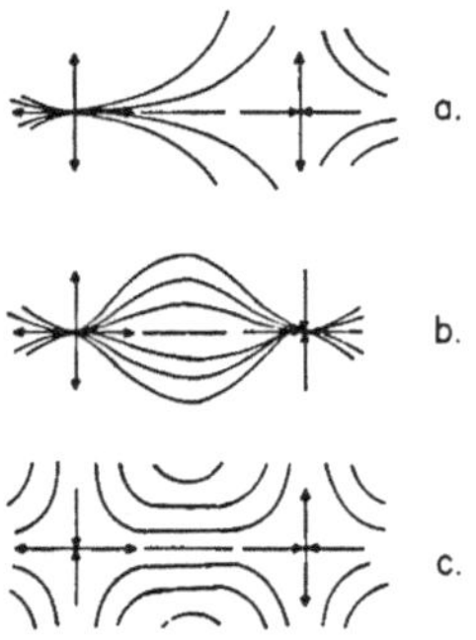


Figure 5.16. Types of transitions between singular points

of the λ's. Configurations b and c do prove to arise in connection with certain types of intermediate shocks to be treated in the next chapter. In the case of fast or slow shocks, however, the loci of extremum points of the curves of these configurations would have to extend out to infinity. This type of behavior is precluded by the requirement that the integral curves in each of the regions formed by the null surfaces must be monotone in configurations space. Hence, if there is a transition between two singular points, the integral curve pattern in the two-space of the two transition eigenvectors, and any other pair of eigenvectors, must be topologically identical to Configuration 5.16a. Adding more dimensions does not change the basic pattern. For example, if positive (or negative) eigenvalues are added at both singular points, a pattern like that of Figure 5.15a is obtained, but if a positive pair is added to one point and a negative pair to the other, the prohibited nonmonotonic pattern would return. In this way, certain configurations of integral curves can be eliminated immediately.

The third concept makes use of the previous two concepts. It is that if a pair of singular points like those of Configuration 5.16a lie adjacent to each other, if integral-curve configurations like those in Figures 5.16b and 5.16c must be excluded by the requirement of monotonic behavior within the various regions, and if there are no singular points in between the indicated pair, then there must be a transition between the two singular points, and the integral curve pattern must be as shown in Figure 5.16a. The best way to be convinced of this is to draw the integral curves in all possible ways. This will indicate that for all integral-curve configurations in which there is no transition between the singular points, the above-stated conditions are violated.

In regard to the fourth concept, note that in order to arrange the integral curves in the "normal" patterns shown in Figures 5.15 and 5.16, the family of eigenvectors at each singular point must be rotated as a whole and then $\vec{\lambda}_1$ with respect to the other four. The normal pattern is quite clearly the one in which the integral curves are subject to minimum bending and twisting; hence, in considering transformations from the actual configuration to the normal one, it is clear that this is not done in an arbitrary way, but along the "path of least resistance."

With the above discussion in mind, we shall reconsider the study of transitions between $SP_3$ and $SP_4$. First, let us examine the integral curves going to $\vec{\lambda}_3$ at $SP_4$ in more detail. In the space of $\vec{\lambda}_1$, $\vec{\lambda}_2$, $\vec{\lambda}_3$; $\lambda_3$ is the corresponding eigenvalue of smallest magnitude; hence, a double infinity of integral curves will arrive at $SP_4$ along $\vec{\lambda}_3$. Tracing these curves backwards, we can see that some may go upward through $F_T = 0$ and some downward through $F_T = 0$, but none through $F_V = 0$. Some will intercept the surface $F_J = 0$, and only then can they go through the surface

$J_z = 0$. The $dT/d\tau$ and $dB_y/d\tau$ of the integral curves which remain in $F_T > 0$, $F_\tau < 0$, $J_z > 0$ are negative; therefore, they head in the general direction of $SP_3$. Since space is everywhere dense with integral curves coming from $\vec{\lambda}_3$ (x decreasing), it is inevitable that some will arrive in the neighborhood of $SP_3$:

Similarly, consider integral curves terminating at $SP_4$ along $\vec{\lambda}_1$ and $\vec{\lambda}_2$. In the plane of these two eigenvectors, $\vec{\lambda}_1$ is the exceptional direction and a singly infinite sheet of integral curves arrive along $\vec{\lambda}_2$. Following them in the direction of decreasing x, we find that they leave $SP_4$ in the region $J_z < 0$, with positive slope $dB_y/d\tau$, and a negative slope $dT/d\tau$. They head in a direction such that they cross the surface $F_J = 0$, whereupon they head back to the plane $J_z = 0$. It is possible that at least one of them terminates at $SP_2$, that is, reaches $J_z = 0$ there. This will be discussed more fully in the next section. It may be possible for some to pass through the surface $J_z = 0$ while they are in $III_2$. If this happens, they acquire a negative slope $dB_y/d\tau$ and head in the direction of $SP_3$. The shape of the surface $F_T = 0$ is such that they could then pass through it and down to the plane $T = 0$, but it is also conceivable that at least one could terminate at $SP_3$.

But study the three-dimensional configuration of integral curves in the eigenvector space of $\vec{\lambda}_1$, $\vec{\lambda}_2$, and $\vec{\lambda}_3$ which is shown in Figure 5.17. The doubly infinite "spray" of integral curves from around $\vec{\lambda}_3$ at $SP_4$ is known to come from $SP_3$. Since $\vec{\lambda}_1$ and $\vec{\lambda}_2$ are the only negative eigenvectors at $SP_3$, it is obvious that they are the only directions from which this spray of integral curves can be brought in from infinity, and also it is known that they do bring in such a spray of integral curves around at least one of the positive eigenvectors. It is also evident that the configuration of Figure 5.17 will be compatible if $\vec{\lambda}_2$ and $\vec{\lambda}_3$ at $SP_4$ are interchanged, although in the latter configuration the curves will be subject to more bending. The compatibility of either of these configurations strongly suggests the existence of a unique slow shock-layer curve; however, it is not possible to determine with certainty at this point whether it goes to $\vec{\lambda}_2$ or to $\vec{\lambda}_3$. Because of the fact that this problem is five-dimensional, and hence impossible to visualize in toto, the discussion of slow shocks will be dropped at this point and renewed in the next chapter in connection with the study of a number of special cases.

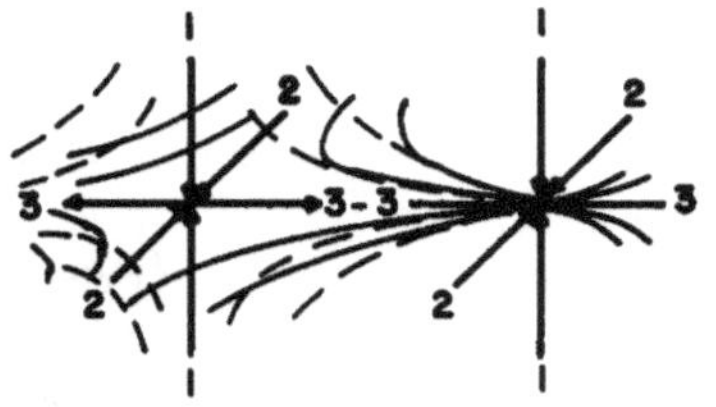

Figure 5.17. A possible configuration of integral curves

## 10. Intermediate Shocks

In the analysis in Chapter 2, based on the conservation laws, intermediate shocks were identified by the condition that the flow velocity upstream of the shock was greater than the normal Alfvén speed, and the velocity downstream was less. In the present context, these are the shocks for which the upstream singular point is either $SP_1$ or $SP_2$ and the downstream point is either $SP_3$ or $SP_4$. From Figure 5.12 and the knowledge which has been obtained on the slopes of integral curves, it is clear that if these transitions exist, they will have to pass through the region $III_2$, and that the slope $dB_y/d\tau$ will have to be largely — if not always — positive. For the same reason, then, $J_z$ will be largely negative.

As a first step in this analysis, consider integral curves in a region $R_5$. We can define $R_5$ as the three-space region $III_2$ in which $J_z$, $F_v < 0$. Equation 5.84 shows that, at least when $B_y < 0$, integral curves cannot enter $R_5$ through $F_T = 0$. Similarly, Equation 5.85 shows that they cannot enter $R_5$ through $F_v = 0$ and Equation 5.86 shows that when $(F_J)_{J_z = 0} < 0$, they cannot enter $R_5$ through $F_T = 0$. Recall that $(F_J)_{J = 0} = 0$ is the projection of $F_J = 0$ onto the space $J_z = 0$, and that in the region of five-space where $J_z < 0$, the null surface $F_J = 0$ lies in between the two branches of the hyperbolic cylinder $(F_J)_{J_z = 0} = 0$; hence, without any restrictions, integral curves cannot enter $R_5$ through $F_T = 0$. Finally, from Equation 5.87, integral curves cannot leave $R_5$ through $J_z = 0$. In addition, from Equations 5.60, 5.62, and 5.63; in $R_5$, $dT/d\tau < 0$ and $dB_y/d\tau > 0$. Note that these slopes are correct for intermediate shock transitions.

Next, let us examine possible transitions from $SP_1$ and $SP_2$ to $\vec{\lambda}_1$ and/or $\vec{\lambda}_2$ at $SP_3$ and $SP_4$. From Inequalities 5.104, these eigenvectors lie in a region which differs from $R_5$ only in the sign of $F_J$. To see where integral curves to these two eigenvectors come from, we shall now trace integral curves from them backwards. A singly infinite sheet of integral curves comes from the $\vec{\lambda}_2$, and judging from the contiguous position of $F_J = 0$, we find that most of them will cross $F_J = 0$ into $R_5$. Then, from the above paragraph, when $B_y < 0$, they can leave $R_5$ ( as x decreaases) only through the surface $J_z = 0$. At the point where $J_z = 0$, $dB_y = 0$; and these curves then acquire a positive slope $dB_y/d\tau$ and thus turn away from $SP_1$ and $SP_2$. Since there is an infinite family of integral curves, some will continue into the region of $R_5$ in which $B_y > 0$. Then they may leave $R_5$ through both $J_z = 0$ and $F_T = 0$. Those which leave through $F_T = 0$ pass through an extremum in $\tau$ and then acquire a positive slope $dT/d\tau$ and a negative slope $dB_y/d\tau$;

but from the form of the null surface $F_{\tau} = 0$, they could still terminate at $SP_1$ or $SP_2$. If they do, however, it is clear that the corresponding shocks will have a short expansion phase before the compressive phase; that is, $\tau$ will not be a monotone function. The curve or curves which remain in $R_5$ also can terminate at $SP_1$ and $SP_2$ if there are eigenvectors at these points in $R_5$. From this discussion, it is clear that it is possible for intermediate shocks to exist if $\vec{\lambda}_3$ or $\vec{\lambda}_4$ at $SP_1$ and/or $SP_2$ lie either in $R_5$ or in the region which differs only in the sign of $F_{\tau}$. In the former case, Inequalities 5.66, 5.67, and 5.70 show that $\lambda < \lambda_{m_2}, \lambda_{\kappa}$ and $P(\lambda) < 0$. In the latter case, $\lambda_{\kappa} < \lambda < \lambda_{m_2}$ and $P(\lambda) < 0$. The discussion of fast shocks makes it clear that some of the integral curves from $\vec{\lambda}_2$ at $SP_1$ could terminate at $SP_3$ or $SP_4$. If they do, these shock-layer curves would differ from those described above in that the function J(x) must have at least one zero.

The above argument shows that intermediate shocks are possible but does not prove their existence. Since the linearized analysis gives no definite information on the whereabouts of $\vec{\lambda}_3$ and $\vec{\lambda}_4$, other than by numerical calculation for special cases, it is not possible to carry through a generally valid proof by actually exhibiting the corresponding shock-layer curve. Because of the difficulties of providing positive proofs of existence of some of the shocks in five-space, we reduce the order of the system in the next chapter by setting various combinations of the dissipation coefficients equal to zero. In these lower (second and third) order systems, we can readily visualize the geometry, and we find that intermediate shocks sometimes exist and sometimes do not exist.

## Chapter 6

# QUALITATIVE STUDY OF THE SHOCK LAYER IN SPECIAL CASES

This chapter has several purposes, achieved for the most part by the solution of certain special cases. One purpose is to examine the effect of current inertia in a simpler context; another—the most important one — is to obtain a better understanding of the properties of intermediate and slow shocks; and a third is to treat two important limiting cases which seem to lie outside the bounds of the general problem of Chapter 5. A final purpose is to assist the reader in following and understanding the full five-dimensional problem by giving examples in which the geometry can be visualized directly.

The first special case treated in this chapter is a two-dimensional problem obtained from Equations 5.13 through 5.17 by setting $m_1 = m_2 = \kappa = 0$. In this context, the effect of current inertia can be studied in the simplest possible manner.

The next three examples are three-dimensional and constitute all of the physically possible cases in which two of the five quantities $m_1$, $m_2$, $\kappa$, $\sigma^{-1}$, $\nu$ are set equal to zero. Thus, $\sigma^{-1}$ and $\nu$ must be kept or discarded together because it has been shown that the current-inertia terms can be of the order of the collision term; and since $\zeta$ and $\eta$ (see page 91) are both positive definite, the condition $m_1 = 0$ requires that $m_2 = 0$. In the first of these examples, only $m_1$, $\sigma^{-1}$, $\nu \neq 0$; in the second, only $m_1$, $m_2$, $\kappa \neq 0$; and in the third, only $\kappa$, $\sigma^{-1}$, $\nu \neq 0$. The first two examples are roughly equal in difficulty; the third is more complicated because there are more special cases to be treated. These are the examples solved primarily to assist in understanding the existence and uniqueness properties of the five-dimensional system. Much can be inferred from these lower-order systems; in particular, the work of Gilbarg[10] and Levinson[61] has provided a great deal of insight into the changes in the integral curves which result if some of the derivative terms are allowed to vanish. As an example, consider the fifth-order system represented by Equations 5.60 through 5.64. If, say, the coefficient $\kappa$ in Equation 5.62 is allowed to become smaller and smaller, the shock-layer curves will move closer and closer to the null surface $F_T = 0$ over as much of their length as the geometrical arrangement of the integral curves will permit. If they cannot lie close to $F_T = 0$ in some region because of slope requirements on the integral curves in five-space, there will be a subshock (a discontinuity) in the limit as $\kappa$ approaches zero; whereas, if they can, the shock-layer curves will be continuous in T as well as in the other variables. In the limit as $\kappa$ approaches zero, Equation 5.62 shows that, away from the null surface $F_T = 0$, all of the five-space integral curves are parallel to the T-axis. This type of parameter variation, in which the variable parameter is the coefficient of one of the highest derivatives, is called a "singular perturbation."

To complete the study, two other important special cases need to be considered. These differ from the above cases in that they are obtained, not by making arbitrary approximations, but by making certain special choices of the parameters. The first of these is a four-dimensional problem defined by the condition $B_x = 0$, and the second is a five-dimensional problem defined by the condition $F_y = 0$. The latter condition produces "switch-on" and "switch-off" shocks.

In the final section of this chapter, the results of the three three-dimensional examples are compared with one another and with the results of the five-dimensional problem. Because of the strong topological similarities between these cases, it has become clear that the nature of the singular points determines the over-all topological behavior of the integral curves to a greater extent than at first appeared possible. An attempt has been made to give these topological similarities a formal mathematical expression.

The analysis of this chapter will rely heavily on the methods and formulas developed in Chapter 5. Hence, the treatment in many places is more abbreviated than would otherwise be possible.

## 1. Negligible Bulk Viscosity, Shear Viscosity, and Thermal Conductivity

Equations 5.13 through 5.17 form the basis for this analysis. If $m_1 = m_2 = \kappa = 0$, then $v$ can be eliminated from Equations 5.15 and 5.17 by means of Equation 5.14, and Equation 5.17 can be simplified with the help of Equation 5.13. After these substitutions have been made, the basic equations for this case become

$$0 = \frac{RT}{\tau} + G^2\tau + \frac{B_y^2}{2\mu} - F_x \tag{6.1}$$

$$0 = \frac{RT}{\gamma - 1} - \frac{G^2\tau^2}{2} + F_x\tau - H + \frac{1}{2G^2}\left(\frac{B_xB_y}{\mu} + F_y\right) - \frac{\tau B_y^2}{2\mu} \tag{6.2}$$

$$\frac{1}{\mu}\frac{dB_y}{dx} = J_z \tag{6.3}$$

$$\nu G\tau^2 \frac{dJ_z}{dx} = -\frac{J_z}{\sigma} + GB_y(\tau - \tau^*) - \frac{B_x F_y}{G} = F_J \tag{6.4}$$

The first two of these equations are identical with Equations 5.20 and 5.21; hence, the intersection of the surfaces defined by these equations is as shown in Figure 5.5. Since the remaining secondary variable, $\tau$, cannot be eliminated, it is logical to consider the problem in the three-space $B_y$-$\tau$-$J_z$. Then the intersection curves of Figure 5.5 form a cylinder in this space—the energy-momentum cylinder—with its generators parallel to the $J_z$-axis. To satisfy the pair of equations 6.1 and 6.2, the integral curves will have to lie on this cylinder.

When $J_z = 0$, the null surface of Equation 6.4, denoted by $F_J$, is identical to Equation 5.22; hence, it reduces to the hyperbolas of Figure 2.2. When $J_z$ increases, the two branches move apart in a linear fashion. Since $\sigma$ is a function of temperature, at least, this null surface is not definitely known; however, $\sigma > 0$, so that its behavior near the singular points is largely governed by $J_z$. The third null surface is, from Equation 6.3, just $J_z = 0$.

Since this problem is basically two-dimensional, the integral curves lie in a two-space, in this case the surface of the energy-momentum cylinder, unless there is a subshock. The qualitative form of the two null curves in this two-space, obtained as the intersections of the surfaces of Equations 6.3 and 6.4 with the energy-momentum cylinder, are shown in Figure 6.1 for case 1 of Figure 5.5. The analysis will be carried through in detail only for this case, although the results of the other two cases will be given.

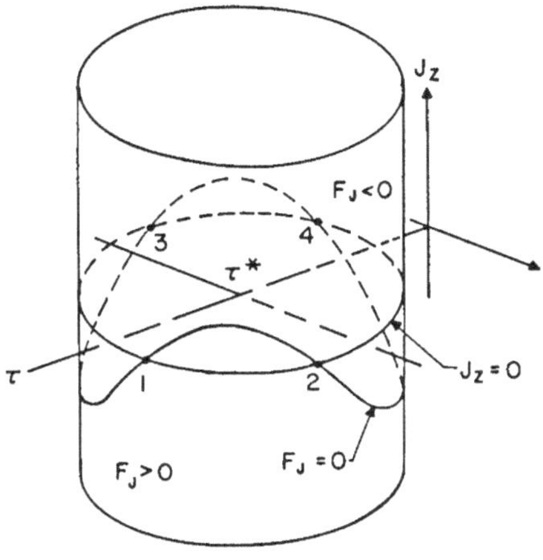


Figure 6.1. Null curves lying on the energy-momentum cylinder

Now that the singular points and null curves have been located, the next step in the analysis is the consideration of the system linearized about the singular points. This linear system is exactly that of Equation 5.25 if $m_1$,

$m_2$, and $\kappa$ are set equal to zero. After doing this, the eigenvector components $\bar{\tau}^*$, $\bar{v}^*$, and $\bar{T}^*$ can be eliminated, with the result

$$\begin{bmatrix} b_x^{\,2}\left[\dfrac{(u^2 - c_f^{\,2})(u^2 - c_s^{\,2})}{u^2(u^2 - a^2)} - \dfrac{\lambda}{\mu\sigma u}\right] & -\dfrac{m_-}{e}\,\dfrac{u^2}{G}\,\dfrac{B_x}{\mu}\,\lambda \\ -\dfrac{m_-}{e}\,\dfrac{u^2}{G}\,\dfrac{B_x}{\mu}\,\lambda & \dfrac{m_-}{m_+}\,u^2 \end{bmatrix} \begin{vmatrix} \bar{B}_y^{\,*} \\ \bar{J}_z^{\,*} \end{vmatrix} = 0 \tag{6.5}$$

In the process, the following equation is also obtained:

$$\bar{\tau}^* = \frac{b_x b_y}{u^2 - a^2}\,\bar{B}_y^{\,*} \tag{6.6}$$

This equation will prove to be a useful adjunct in understanding the behavior of the integral curves in the three-space of Figure 6.1.

As usual, we can obtain the eigenvalues by setting the determinant of the linear system equal to zero. With the help of Equations 4.62 and 4.64, the determinant of Equation 6.5 reduces to

$$\lambda^2 + \frac{\nu_c}{u}\lambda - \frac{\omega_p^{\,2}}{c^2}\,\frac{(u^2 - c_f^{\,2})(u^2 - c_s^{\,2})}{u^2(u^2 - a^2)} = 0 \tag{6.7}$$

which has an interesting resemblance to Equation 5.9 for the eigenvalues of the linear system in the normal electric field and conduction current. Comparing the two, we see that both effects are damped by collisions, but the wave lengths of the spatial oscillations — if they exist — are shorter for the electric-field effect than for the current-inertia effect in the ratio, roughly, of $u/c$. Even though $c^2$ appears in the denominator of the constant term in Equation 6.7, $\omega_p^{\,2}$ can be large enough to make this term important.

The solution of Equation 6.7 is

$$\lambda = \frac{\nu_c}{2u}\left[-1 \pm \sqrt{1 + \frac{4\,\omega_p^{\,2}}{\nu_c^{\,2}}\,\frac{(u^2 - c_f^{\,2})(u^2 - c_s^{\,2})}{c^2(u^2 \quad a^2)}}\right] \tag{6.8}$$

Then, with the help of Inequalities 2.51, the character of the eigenvalues at each of the four singular points is as follows:

$$SP_1,\ SP_{2,3}\ (u < a): \quad \lambda_1 < 0 < \lambda_2, \qquad \text{both real}$$

$$SP_{2,3}\ (u > a),\ SP_4: \ \lambda_1 < \lambda_2 < 0 \text{ if real}, \quad \mathrm{Re}(\lambda_1, \lambda_2) < 0 \text{ if complex}$$

As to the eigenvectors, the second of Equations 6.5 shows that $\overline{B}_y^*$ and $\overline{J}_z^*$ have the same (opposite) sign if $\lambda$ is positive (negative). We can find the sign of $\overline{\tau}^*$ from Equation 6.6 by recalling that $b_y > 0$ at $SP_1$ and $SP_2$, and $b_y < 0$ at $SP_3$ and $SP_4$. At $SP_2$, if $u > a$, $\overline{\tau}^*$ and $\overline{B}_y^*$ have opposite signs. This means that $SP_2$ lies to the high-$\tau$ side of the point on the cylinder where $dB_y = 0$. If $u < a$, the opposite is true. Similar conditions are found at $SP_3$. When $u < a$, $SP_3$ lies to the low-$\tau$ side of $dB_y = 0$, and vice versa when $u > a$. Correspondingly, it is evident that the slope of the projections of the integral curves at $SP_1$ and $SP_4$ onto the $B_y$-$\tau$ plane is always negative. Realization of this fact will be important in the discussion of intermediate shocks.

Away from the singular points, the slope of the integral curves is found by dividing Equation 6.4 by Equation 6.3. Thus,

$$\frac{dJ_z}{dB_y} = \frac{1}{\mu \nu G \tau^2} \frac{F_J}{J_z} \tag{6.9}$$

The direction in which the integral curves cross the null curve $J_z = 0$ can easily be found from Equation 6.4. It shows that $dJ_z/dx$ has the same sign as $F_J$.

Consider transitions from $SP_1$ to $SP_2$ when $u > a$ at $SP_2$. The projection of the null curves and integral curves in the region of these singular points onto the $J_z$-$B_y$ plane is as shown in Figure 6.2. $SP_1$ is a saddle point, $SP_2$ is a downstream node, and $\vec{\lambda}_1$ and $\vec{\lambda}_2$ lie in the relative positions shown. Furthermore, since $|\lambda_1| > |\lambda_2|$ at $SP_2$, an infinite family of integral curves come in to $SP_2$ in the direction of $\vec{\lambda}_2$. Because of the pattern of integral curves that appears in Figure 6.2 and the fact that this pattern also holds for cases 2 and 3 of Figure 5.5, it is obvious that a unique fast-shock layer exists when $u > a$.

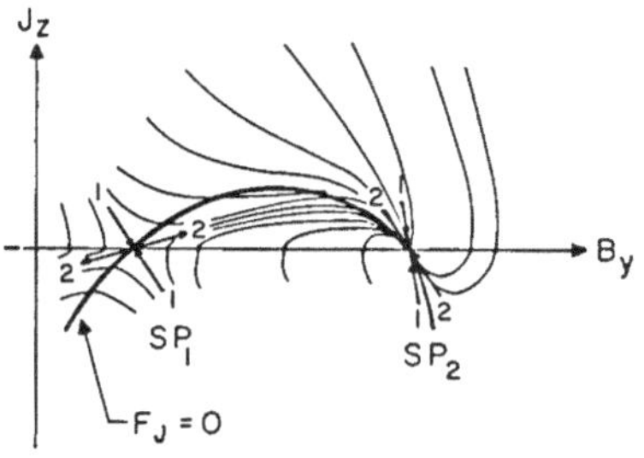


Figure 6.2. Integral curve pattern showing the existence of a unique fast shock

Consider transitions from $SP_3$ to $SP_4$ when $u < a$ at $SP_3$. A little study will show that the pattern of integral curves is exactly as shown in Figure 6.2 when both slow-shock singular points exist; therefore, a unique slow shock exists if $u < a$.

Consider transitions from $SP_1$ to $SP_3$ when $u > a$ at $SP_3$. For this case, the pattern of integral curves projected in the $J_z - B_y$ plane is as shown in Figure 6.3 for all cases of Figure 5.5. When the available knowledge of the behavior of the integral curves is applied, it may be seen that they take a form which must result in a unique transition between $SP_1$ and $SP_3$.

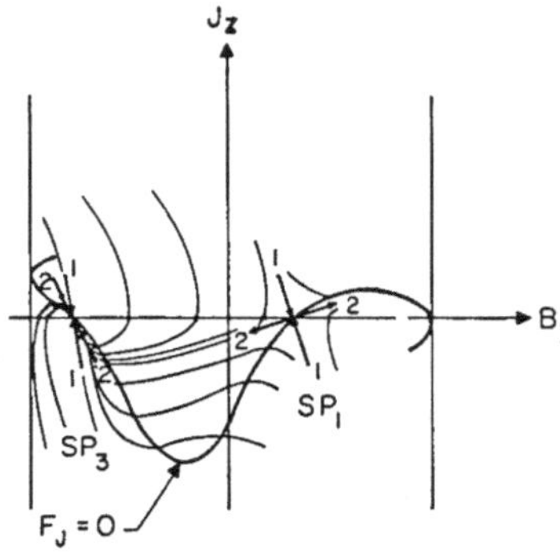


Figure 6.3. Integral curves showing the existence of a unique intermediate $(1 \rightarrow 3)$ shock

When $u < a$ at $SP_2$ (possible only in case 1, Figure 5.5), the pattern of integral curves on the low-$\tau$ side of the energy-momentum cylinder is topologically identical to that of Figure 6.3; hence, there is a unique transition from $SP_2$ to $SP_4$ in case 1 when $u < a$ at $SP_2$. The fact that the transition goes to $SP_4$ clearly precludes the possibility of a transition to $SP_3$; thus, there is no $2 \rightarrow 3$ shock.

The fact that $\overline{\tau}^*$ and $\overline{B}_y^*$ have opposite signs at $SP_1$ and like signs at $SP_3$ when $u > a$ means that the $1 \rightarrow 3$ shock is first expansive, then compressive. For the same reason, the $2 \rightarrow 4$ shock is first compressive, then expansive. The total entropy rise across both of these shocks, however, is still positive. It should be recalled from Chapter 3 that both of these shocks are unstable with respect to small disturbances.

All four of the above transitions have the common characteristic that they never cross either of the two generators of the energy-momentum cylinder for which $dB_y = 0$ (this point clearly exists only in case 1). Study of Figures 6.2 and 6.3 will show that if there were any transitions crossing $dB_y = 0$, the conditions on the slopes of the integral curves are such that they would have to cross at a point other than $J_z = 0$. But then Equation 6.3 shows that $dx$ must vanish as $dB_y$ vanishes. This means that the point $dB_y = 0$ is an extremum point for $x$. Consequently, $x$ does not increase monotonically across the line $dB_y = 0$, with the result that no transition curves for which this occurs can correspond to shocks. The only way for transitions to exist in this case is by the appearance of a subshock.

The results of this special case are summarized in Table 6.1. On the left are the possible initiating points for shocks, characterized by the fact that there is a positive eigenvalue at each of

Table 6.1

Unique Transitions in the Special Case where $m_1 = m_2 = \kappa = 0$

| | **$SP_2$ u > a** | **$SP_2$ u < a** | **$SP_3$ u > a** | **$SP_3$ u < a** | $SP_4$ |
|---|---|---|---|---|---|
| $SP_1$ | 1 → 2 | subshock | 1 → 3 | **subshock** | * |
| **$SP_2$ (u < a)** | — | — | — | * | 2 → 4 |
| **$SP_3$ (u < a)** | x | x | — | — | 3 → 4 |

them. On the top the possible terminating points for shocks are listed. The four cases for which unique shock-layer curves exist and for which there is no subshock are indicated by appropriate transition symbols. It should be noted that the 1 → 2 and 1 → 3 shocks are supersonic both ahead and behind, whereas the 2 → 4 and 3 → 4 shocks are subsonic both ahead and behind. Moreover, each of the terminating singular points for these four shocks has either two negative real eigenvalues or a pair of complex-conjugate eigenvalues. (The criterion as to which occurs is obvious from Equation 6.8.) The transitions in Table 6.1 designated "subshock" are separated by one of the lines $dB_y = 0$, those indicated by x's will violate the second law of thermodynamics, and the starred transitions do not exist because the two corresponding singular points are separated by a third singular point on the energy-momentum surface. Note that in cases 2 and 3 of Figure 5.5, only $SP_1$ can initiate shock transitions.

Summarizing, we can say that the fast shock is the only one which has been shown always to exist, and when u < a at $SP_2$, there is a subshock. The slow shock exists if u < a at $SP_3$. Unique 2 → 4 and 3 → 4 intermediate shocks sometimes exist, but 1 → 4 and 2 → 3 intermediate shocks never do. Since the analysis of the five-dimensional system strongly suggested the existence of a slow shock, it is apparently true that the eigenvector at $SP_3$ to which it corresponds is one of those that have disappeared in the reduction to the present two-dimensional system. Hence, it is likely that the slow shock also always exists, but that it contains a subshock when u > a at $SP_3$.

## 2. Negligible Shear Viscosity and Thermal Conductivity

When $m_2 = \kappa = 0$, and v is eliminated by means of Equation 5.14, Equations 5.13, 5.15, 5.16, and 5.17 become

$$m_1 G \frac{d\tau}{dx} = \frac{RT}{\tau} + G^2 \tau + \frac{B_y^2}{2\mu} - F_x = F_\tau \quad (6.10)$$

$$0 = \frac{RT}{\gamma - 1} - \frac{G^2\tau^2}{2} + F_x \tau - H + \frac{1}{2G^2}\left(\frac{B_x B_y}{\mu} + F_y\right)^2 - \frac{\tau B_y^2}{2\mu} = F_T \quad (6.11)$$

$$\frac{1}{\mu}\frac{dB_y}{dx} = J_z \tag{6.12}$$

$$\nu G\tau^2 \frac{dJ_z}{dx} = -\frac{\nu\tau}{m_1} F_\tau J_z - \frac{J_z}{\sigma} + GB_y(\tau - \tau^*) - \frac{B_x F_y}{G} = F_J \tag{6.13}$$

Although some of these equations have appeared before, they are rewritten and renumbered here for convenience of reference.

In this problem, the primary variables are $\tau$, $B_y$, and $J_z$, and there is a secondary variable $T$, which can be completely eliminated, as in the previous example. Consequently, for any shock-layer curves which appear in $\tau$-$B_y$-$J_z$ space, $\tau$, $B_y$, and $J_z$ will be continuous functions of $x$. Equation 6.11 shows, then, that $T$ will also be a continuous function of $x$, with the result that there will be no subshock if $m_1$, $\nu$, and $\sigma^{-1}$ are all different from zero.

Null Surfaces. The null surfaces can be most conveniently represented in the space of the principal variables, in this case the space of Figure 6.1. Comparison of the above equations with those of Section 1 makes it clear that the energy-momentum cylinder shown in Figure 6.1 is one of the null surfaces — the one on which $F_\tau = 0$. In this problem, however, the integral curves are not constrained to lie on this cylinder, because $m_1 \neq 0$. The next null surface is again simply $J_z = 0$; and the final one, $F_J = 0$, can be visualized sufficiently well if it is written in the form

$$GB_y(\tau - \tau^*) = \frac{B_x F_y}{G} + \left(\frac{1}{\sigma} + \frac{\nu\tau}{m_1} F_\tau\right) J_z$$

First, its intersection with the energy-momentum cylinder of Figure 6.1 is still as shown in Figure 6.1, because when $F_\tau = 0$, the above equation is identical with $F_J = 0$ in Equation 6.4. Second, when $J_z = 0$, $F_J$ reduces to the hyperbola shown in Figure 5.12. Finally, when $\tau = \tau^*$ and $B_y = 0$, Equation 6.10 shows that $F_\tau$ has some definite negative value, but is not known whether $1/\sigma \gtrless \nu\tau|F_\tau|/m_1$. Hence, the intersection occurs either above or below the plane $J_z = 0$, but in any case $F_J = 0$ must be a continuous surface bridging the space between the four singular points.

Properties of the Linearized System. If Equations 6.10 through 6.13 are linearized about a singular point, and the secondary variable $T$ is eliminated, the result is exactly that obtained from Equation 5.25 by setting $m_2 = \kappa = 0$ and eliminating $\overline{v}^*$ and $\overline{T}^*$. After doing this, the following third-order system results:

$$\begin{bmatrix} u^2 - a^2 - \dfrac{m_1 u^2}{G}\lambda & b_x b_y & 0 \\ b_x b_y & b_x^{\,2}\left(\dfrac{u^2 - b_x^{\,2}}{u^2} - \dfrac{\lambda}{\mu\sigma u}\right) & -\left(\dfrac{m_-}{e}\,\dfrac{u^2}{G}\,\dfrac{B_x}{\mu}\right)\lambda \\ 0 & -\left(\dfrac{m_-}{e}\,\dfrac{u^2}{G}\,\dfrac{B_x}{\mu}\right)\lambda & \dfrac{m_-}{m_+}\,u^2 \end{bmatrix} \begin{Vmatrix} \bar{\tau}^* \\ \bar{B}_y^{\,*} \\ \bar{J}_z^{\,*} \end{Vmatrix} = 0 \tag{6.14}$$

If the notation $D_2(\lambda)$ is used for the determinant formed by the first two rows and columns of Equations 6.14, the condition that the determinant of Equations 6.14 vanishes can be written in the form

$$D_2(\lambda) + c\,\lambda^2(\lambda - \lambda_{m_1}) = 0 \tag{6.15}$$

in which $c$ is a positive constant, and

$$\lambda_{m_1} = \frac{G}{m_1 u^2}\,(u^2 - a^2) \tag{6.16}$$

The signs of the real eigenvalues of $D_2(\lambda) = 0$ can be most easily found by applying the "Inertial Theorem for Quadratic Forms" (page 133). Thus, the quadratic form of the part of $D_2(\lambda)$ not proportional to $\lambda$ is

$$V_{ij}\, a_i\, a_j = (u^2 - a^2)\,\bar{\tau}^{*2} + 2 b_x b_y\, \bar{\tau}^*\bar{B}_y^{\,*} + \frac{b_x^{\,2}}{u^2}\,(u^2 - b_x^{\,2})\,\bar{B}_y^{\,*2}$$

$$= (u^2 - a^2)\left(\bar{\tau}^* + \frac{b_x b_y}{u^2 - a^2}\,\bar{B}_y^{\,*}\right)^2 + \frac{b_x^{\,2}(u^2 - c_f^{\,2})(u^2 - c_s^{\,2})}{u^2(u^2 - a^2)}\,\bar{B}_y^{\,*2} \tag{6.17}$$

from which the signs of $D_2(\lambda) = 0$ can be tabulated as follows:

$SP_1$: $0 < \bar{\lambda}_1 < \bar{\lambda}_2$

$SP_{2,3}$: $\bar{\lambda}_1 < 0 < \bar{\lambda}_2$

$SP_4$: $\bar{\lambda}_1 < \bar{\lambda}_2 < 0$

**With these results, the graphical procedure of Section 5 in Chapter 5 can be applied to Equation 6. 15 to show that the eigenvalues of Equations 6. 14 have the following properties:**

$$\left.\begin{array}{ll} SP_1: & \lambda_1 < 0 < \lambda_2 < \lambda_{m_1} < \lambda_3 \quad , \quad \text{all real} \\ SP_{2,3}: & \lambda_1 < \lambda_2 < \left\{\begin{array}{l} 0 < \lambda_{m_1} \\ \lambda_{m_1} < 0 \end{array}\right\} < \lambda_3 \quad , \quad Re(\lambda_1, \lambda_2) < 0 \\ SP_4: & \lambda_1 < \lambda_2 < \lambda_{m_1} < \lambda_3 < 0 \quad , \quad Re(\lambda_1, \lambda_2) < 0 \end{array}\right\} \quad (6.18)$$

**At $SP_{2,3,4}$, $\lambda_1$ and $\lambda_2$ may be either real or complex; all the other eigenvalues are real.**

**The directions of the eigenvectors can easily be found from the first and third of Equations 6. 14. Thus,**

$$\bar{B}_y^* \propto \frac{\lambda - \lambda_{m_1}}{B_y} \bar{\tau}^* \qquad (6.19)$$

$$\bar{J}_z^* \propto \lambda \bar{B}_y^* \qquad (6.20)$$

General Properties Related to the Integral Curves. To see the behavior of the integral curves away from the singular points, the following proportionalities, obtained from Equations 6. 10, 6. 12, and 6. 13 will be useful:

$$\frac{dJ_z}{dB_y} \propto \frac{F_J}{J_z} \; , \; \frac{dJ_z}{d\tau} \propto \frac{F_J}{F_\tau} \; , \; \frac{dB_y}{d\tau} \propto \frac{J_z}{F_\tau} \qquad (6.21)$$

The directions in which the integral curves cross the null surfaces $F_\tau = 0$ and $J_z = 0$ can easily be found. From Equation 6. 10,

$$\left.\frac{dF_\tau}{dx}\right]_{F_\tau=0} = \frac{R}{\tau} \left.\frac{dT}{dx}\right]_{F_\tau=0} + B_y J_z$$

Then from Equation 6. 11,

$$\frac{R}{\gamma - 1} \left.\frac{dT}{dx}\right]_{F_\tau=0} = \left[GB_y(\tau - \tau^*) - \frac{F_y B_x}{G}\right] \frac{J_z}{G} = \frac{J_z}{G} (F_J)_{J_z=0}$$

the latter definition coming from Equation 6.13. Combining the above two expressions, we have

$$\left.\frac{dF_\tau}{dx}\right]_{F_\tau=0} = \left[\frac{(\gamma - 1)}{\tau G}(F_J)_{J_z=0} + B_y\right] J_z \tag{6.22}$$

The direction along which integral curves cross $J_z = 0$ can be obtained directly from Equation 6.13, and, as in the five-dimensional problem, the direction along which curves cross $F_J = 0$ is indefinite.

Another useful piece of information is the slope of the null surfaces at the singular points. The slope in the $B_y - \tau$ plane of the energy-momentum cylinder, given on page 122, is particularly helpful in establishing the behavior of the integral curves. Since the bracketed term is zero at the singular points,

$$\frac{dB_y}{d\tau} = - \frac{\mu(u^2 - a^2)}{B_y \tau^2} \tag{6.23}$$

Fast Shocks. With the above information available, it is possible to look for transitions between singular points. First consider the region around $SP_1$ and $SP_2$. By using Equations 6.19 and 6.20, one can deduce that the eigenvectors lie, as shown in Figure 6.4, relative to the null surfaces. This figure is a small portion of Figure 6.1. Note that it is valid for all three of the cases of Figure 5.5. Although $SP_1$ is a saddle point in the planes $\vec{\lambda}_1, \vec{\lambda}_2$ and $\vec{\lambda}_1, \vec{\lambda}_3$; it is an upstream node in the plane $\vec{\lambda}_2, \vec{\lambda}_3$, in which $\vec{\lambda}_3$ is the exceptional direction. On the other hand, $SP_2$ is a saddle point in planes $\vec{\lambda}_1, \vec{\lambda}_3$ and $\vec{\lambda}_2, \vec{\lambda}_3$ and a downstream node in plane $\vec{\lambda}_1, \vec{\lambda}_2$, in which $\vec{\lambda}_1$ is the exceptional direction.

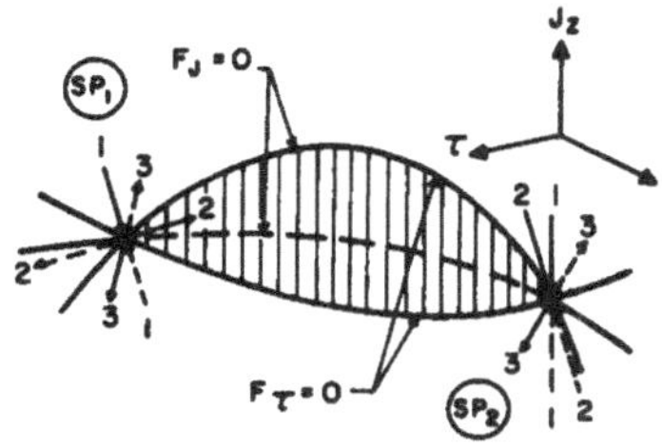


Figure 6.4. Eigenvectors at $SP_1$ and $SP_2$. Solid vectors lie above $\tau$-$B_y$ plane.

The two eigenvectors $\vec{\lambda}_3$ of Figure 6.4 point towards the reader into the region $F_\tau > 0$, $F_J > 0$, $J_z > 0$. Since all three slopes of Equation 6.21 are positive in this region, it is clear that integral curves from these two eigenvectors go off to infinity and cannot correspond to shocks. The $\vec{\lambda}_3$ also point in the opposite direction, towards $SP_3$ and $SP_4$, and may originate intermediate shocks, which will be discussed later in this section. Equation 6.22 shows that on the segment of the energy-momentum cylinder between $SP_1$ and $SP_2$, integral curves move from $F_T < 0$ to $F_T > 0$ (outward) when $J_z > 0$, and inward when $J_z < 0$. When both $J_z = 0$ and $F_T = 0$, they are parallel to the $J_z$-axis and move from $J_z < 0$ to $J_z > 0$. Thus, a picture is obtained of integral curves starting from infinity in the region $F_T > 0$, $F_J > 0$, $J_z < 0$ and moving upward toward the energy-momentum cylinder ($F_T = 0$). They turn around at $J_z = 0$ and return to infinity within the region $F_T > 0$, $F_J > 0$, $J_z > 0$, some intersecting $F_T = 0$ twice, some tangent to it at $J_z = 0$, and some not making the intersection at all. This behavior is totally consistent with the known behavior of the integral curves near the singular points.

Within the closed region $F_T < 0$, $F_J > 0$, $J_z > 0$, Equations 6.21 show that the slopes $dJ_z/d\tau$ and $dB_y/d\tau$ are negative, and $dJ_z/dB_y$ is positive; hence, the single infinity of curves coming from $\vec{\lambda}_2$ at $SP_1$ move toward $SP_2$. They must eventually intersect $F_J = 0$, whereupon they move toward $J_z = 0$, that is, toward $SP_2$. It is clear from these arguments that the configuration of integral curves around $SP_1$ and $SP_2$ is topologically identical to that of Figure 5.15a; consequently, a unique shock-layer curve connects $SP_1$ to $SP_2$. From Inequalities 6.18 it is evident that $\lambda_1$ and $\lambda_2$ can be complex at $SP_2$, thus causing it to be a spiral point in the plane of $\vec{\lambda}_1$ and $\vec{\lambda}_2$. The signs of the slopes and directions in which they can cross the null surfaces remain the same, however, thus resulting in the same conclusion.

Slow Shocks. Now consider the region around $SP_3$ and $SP_4$. From Equations 6.19 and 6.20, the eigenvectors lie relative to the null surfaces as shown in Figure 6.5. The energy-momentum null surface is shown in this figure as it appears in case 1, Figure 5.5. Here $SP_3$ is a saddle point in the $\vec{\lambda}_1$, $\vec{\lambda}_3$ and $\vec{\lambda}_2$, $\vec{\lambda}_3$ planes; therefore, since $\lambda_3$ is the only positive eigenvalue, it is clear that if a transition to $SP_4$ exists, it must be unique, for there is only one integral curve coming from $SP_3$ in the direction of $\vec{\lambda}_3$.

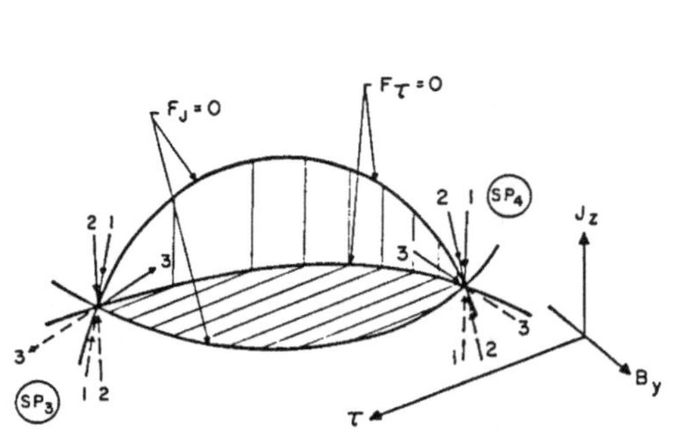

Figure 6.5. Eigenvectors at $SP_3$ and $SP_4$. Solid vectors lie above $\tau$-$B_y$ plane.

In all three eigenvector planes, $SP_4$ is a downstream node. Since $\lambda_3$ has the smallest magnitude there (see Inequalities 6.18), $\vec{\lambda}_3$ is the only eigenvector which receives an infinite number of integral curves in all directions around it, i.e., a double infinity of curves. If one takes into account the pattern of integral curves around $SP_3$, this fact strongly suggests a unique transition between the two $\vec{\lambda}_3$'s; however, it does not prove it. To disprove this conjecture, on the other hand, it would be necessary to show how the integral curves could bypass $SP_4$ without the indicated intersection. Suppose they did bypass $SP_4$. Then, on the other side of $SP_4$, the integral curves moving (as x increases) away from $SP_3$ pass integral curves approaching $SP_4$. But this is impossible according to an argument given on page 162 (first topological concept). Consequently, when an adjacent pair of singular points like $SP_3$ and $SP_4$ exist, the curves from the region around the upstream point ($SP_3$) must terminate at the downstream point ($SP_4$); hence, a unique slow shock exists. Consideration of the slopes of the integral curves in the regions around $SP_3$ and $SP_4$, and the direction and slope with which they cross the null surfaces, brings further strength and understanding to this conclusion.

To see that the transition must go to $\vec{\lambda}_3$ at $SP_4$, it is only necessary to examine the signs of the slopes of the eigenvectors and of the integral curves in the various regions. At $SP_4$ the important branch of $\vec{\lambda}_3$ lies in the region $F_\tau < 0$, $F_J < 0$, $J_z > 0$, which is correct for a transition from $SP_3$. The eigenvectors $\vec{\lambda}_1$ and $\vec{\lambda}_2$, on the other hand, lie in $F_\tau > 0$, $F_J < 0$, $J_z > 0$, where the slopes are such that integral curves can only come to them from infinity, or in $F_\tau < 0$, $F_J > 0$, $J_z < 0$, where transitions can only come from $SP_1$ and/or $SP_2$. Thus, $\vec{\lambda}_3$ is the terminating eigenvector. Since $\vec{\lambda}_3$ is always real, there can be no spatial oscillations downstream of a slow shock in this case.

Intermediate Shocks. For the analysis of intermediate shocks, we shall examine the region for which $F_\tau < 0$, $F_J \Big|_{J_z=0} < 0$.

First, we observed in the study of fast shocks that $\vec{\lambda}_3$ at $SP_1$ and $SP_2$ point into the region $F_\tau < 0$, $F_J < 0$, $J_z < 0$; and in the study of slow shocks that $\vec{\lambda}_1$ and $\vec{\lambda}_2$ at $SP_3$ and $SP_4$ both point toward these singular points from the region $F_\tau < 0$, $F_J > 0$, $J_z < 0$. If there are intermediate shocks, they will have to originate and terminate in these eigenvectors. From Equation 6.21, it can be seen that $dB_y/d\tau > 0$ in this region, and this is the correct sign for transitions from $SP_1$ and $SP_2$ to $SP_3$ and $SP_4$.

When $m_1$ becomes infinitesimally small, Equation 6.10 shows that all integral curves a finite distance from the null surface $F_\tau = 0$ must be very nearly parallel to the $\tau$-axis, and only at points for which $F_\tau$ is infinitesimal can the curves depart appreciably from the direction of the $\tau$-axis. This is very nearly the case of the previous example. Now, however, there is a third eigenvector, $\vec{\lambda}_3$, which is very nearly parallel to the $\tau$-axis when $m_1$ is very small. The integral curve from $\vec{\lambda}_3$ at $SP_2$ will go straight over to the neighborhood of the surface $F_T = 0$ (subshock) and follow very close to that surface down to $SP_4$. In cases 2 and 3 of Figure 5. 5, this implies that there will be a compression followed by an expansion phase within the shock layer.

The only configuration of the integral curve which satisfies the above-described behavior is shown topologically in Figure 6. 6a. Hence, when $m_1$ is small compared with $\nu$ and $\sigma^{-1}$, unique $1 \rightarrow 3$ and $2 \rightarrow 4$ shocks exist, an infinite family of $1 \rightarrow 4$ shocks exists, and there is no $2 \rightarrow 3$ shock.

The eigenvector $\lambda_1$ is an exceptional direction at every singular point; hence, in Figure 6. 6a one can visualize four straight lines perpendicular to the plane of the singular points going upwards and downwards to infinity. Since $\lambda_1$ is the only negative

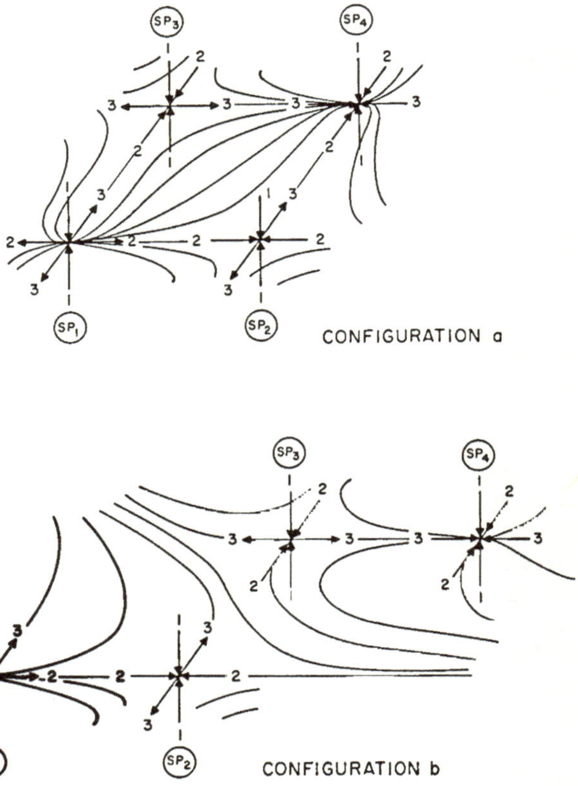


Figure 6. 6. A summary of all possible transitions between singular points for the case $m_2 = \kappa = 0$

eigenvalue at $SP_1$, there is a cylindrically symmetrical "horn" of integral curves coming down from infinity above and a similar horn coming up from infinity below. A singly infinite set of hyperbolas lying in the $\vec{\lambda}_1$, $\vec{\lambda}_2$ plane to the right of $SP_1$ terminate at $SP_2$; whereas those in the similar $\vec{\lambda}_1$, $\vec{\lambda}_3$ plane between $SP_1$ and $SP_3$ terminate at $SP_3$. All of the curves in the space between these two planes and inside the infinite rectangular box terminate at $SP_4$. A similar analysis shows that the required behavior of the integral curves in the other two planes of the box is also compatible with Figure 6. 6a.

If the current-inertia effect is allowed to vanish, $\vec{\lambda}_1$ becomes infinitely long, and the whole three-dimensional pattern of integral curves collapses into the plane of the singular points shown in Figure 6. 6a. The truth of this statement can be seen from Equation 6. 13. When $F_J \neq 0$, $\nu \rightarrow 0$ means that $dJ_z \rightarrow \infty$; therefore, at any particular point in $B_y$-$\tau$-$J_z$ space at which $F_J \neq 0$, the integral curves become more and more nearly parallel to the $J_z$-axis. This also shows that the integral curves collapse into the plane $F_J = 0$ when $\nu \rightarrow 0$, as they should.

From the formula for $dB_y/d\tau$ (Equation 6. 21), from the known positions of the eigenvectors, and by referring to Figure 5. 12, it is clear that $J_z < 0$ along the unique transitions $1 \rightarrow 3$, $2 \rightarrow 4$. For the infinity of transitions $1 \rightarrow 4$, however, the positions of the eigenvector $\vec{\lambda}_2$ at $SP_1$ and $\vec{\lambda}_3$ at $SP_4$ indicate that $dB_y$ must vanish twice along its path. This means that $F_J(dJ_z/dx)$ must vanish three times, with the result that the current profile is as shown in Figure 5. 13. Since $F_J = 0$ is a continuous null surface containing the four singular points, it is entirely possible for the indicated family of transitions to cross it three times.

The configuration of integral curves which results when $\lambda_1$ and $\lambda_2$ are complex remains to be discussed. The singular points $SP_{2,3,4}$ are then downstream spiral points in planes defined by two vectors: $\mathrm{Re}(\vec{\lambda}_1)$ and $\mathrm{Im}(\vec{\lambda}_1)$. The projections of the integral curves in any plane close to and parallel to this one must also be a spiral; however, the required integral-curve slopes are such that these spirals "corkskrew" away from the plane through the singular points. Thus, in the vertical plane of Figure 6. 6a passing through $SP_1$ and $SP_2$ there is a singly infinite family of spirals coming into $SP_2$, only one of which comes from $SP_1$. Curves coming down from the "horn" around $\vec{\lambda}_1$ at $SP_1$ and inside the square must spiral around $\vec{\lambda}_3$ at $SP_3$, but these spirals must move toward $SP_4$ and intersect it along $\vec{\lambda}_3$. Similar considerations at the other singular points show that for all the other transitions, the fact of complex eigenvalues does not change the existence and uniqueness properties of shocks.

Now consider the case for which $m_1$ is large compared with

$\nu$ and $\sigma^{-1}$. Equation 6.10 shows that at a given point $F_\tau$ in configuration space, the larger $m_1$ becomes the smaller is $d\tau$ for a given $dx$. Thus, the curves become more and more nearly perpendicular to the $\tau$-axis. As $\sigma^{-1}$ and $\nu$ become infinitesimal, the integral curves move closer and closer to the null surface $F_J = 0$. Meanwhile, the eigenvectors $\vec{\lambda}_3$ at $SP_1$ and $SP_2$ rotate counterclockwise until, in the limit as $\nu/m_1$ and $\sigma^{-1}/m_1$ vanish, they become perpendicular to the $\tau$-axis. In this circumstance, Configuration a of Figure 6.6 must have switched over to Configuration b. A little study will show that this new configuration is completely compatible with all of the requirements for integral curves, and also that it passes smoothly into Configuration a as the above ratios increase. In Configuration b, there are no intermediate shocks, but it is clear that there must be a unique set of ratios $\nu/m_1$ and $\sigma^{-1}/m_1$ for which a $2 \rightarrow 3$ shock-layer curve exists. This must occur for a rather small ratio $\sigma^{-1}/m_1$ because the resistivity must be small to permit a current $J_z$ large enough to give the large change in magnetic field from $SP_2$ to $SP_3$, and the first viscosity, $m_1$, must be large to prevent a large change in specific volume $\tau$. For this reason, one can say in this example that intermediate shocks will usually exist.

These results agree with those of Germain[13] and Kulikovskii and Liubimov[45] for the case in which current inertia is neglected; however, in the latter the final conclusions must have been inadvertently reversed. Mainly from the topological considerations, it is now clear that the addition of current inertia cannot change the qualitative results on existence and uniqueness; it can only change the shape of the shock-layer curves.

## 3. Negligible Electrical Conductivity and Current Inertia

The current-inertia terms are those terms of Equation 5.17 proportional to $\nu$. When $\nu$ and the electrical resistivity $\sigma^{-1}$ are set equal to zero, the remaining expression, $G\tau B_y - vB_x = 0$, can be used to eliminate $B_y$ from Equations 5.13, 5.14, and 5.15. Since $J_z$ does not appear in these equations, Equation 5.16 is not needed. The equations of this special case are, therefore,

$$m_1 G \frac{d\tau}{dx} = \frac{RT}{\tau} + G^2\tau + \frac{\tau^*}{2}\frac{v^2}{\tau^2} - F_x = F_\tau \qquad (6.24)$$

$$m_2 \frac{dv}{dx} = \left(1 - \frac{\tau^*}{\tau}\right) Gv - F_y = F_v \qquad (6.25)$$

$$\frac{\kappa}{G}\frac{dT}{dx} = \frac{RT}{\gamma - 1} - \frac{G^2\tau^2}{2} + F_x\tau - H - \frac{1}{2}\left(1 - \frac{\tau^*}{\tau}\right)v^2 + \frac{F_y}{G} v = F_T \qquad (6.26)$$

Null Surfaces. First, consider $F_\tau = 0$. From Equation 6.24, one may write

$$RT = -G^2\tau^2 + F_x\tau - \frac{\tau^*}{2}\frac{v^2}{\tau} \qquad (6.27)$$

Discussion of this equation should be compared with the discussion on pages 8-9. For constant $\tau$, this is a family of inverted parabolas symmetric about $v = 0$. For constant $v$, it is parabolic when $\tau$ is large and hyperbolic when $\tau$ is small. The intersection with the plane $T = 0$ is now a continuous closed curve intersecting the $\tau$-axis at $\tau = 0$, $F_x/G^2$; and the point of maximum temperature of the surface is at exactly the same place as in Figure 2.1. This null surface is shown in Figure 6.7, in which the various maxima are given. A particular feature to note is that at the origin, the slope in the $v$-$\tau$ plane is discontinuous.

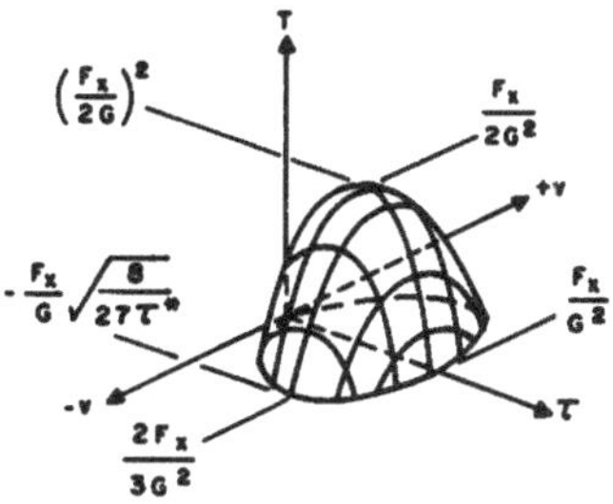

Figure 6.7. The normal-momentum null surface ($F_\tau = 0$)

Second, consider $F_v = 0$. From Equation 6.25 this equation may be written

$$(\tau - \tau^*)\,v = \frac{F_y}{G}\,\tau \qquad (6.28)$$

Comparing Equation 2.9, we can see that $v$ becomes unbounded in both equations at $\tau = \tau^*$; and if we choose $F_y$ to be positive as before, when $\tau > \tau^*$, $v > 0$. The differences between the two equations are that now $v \to \text{const} > 0$ as $\tau \to \infty$, and $v \to 0$ as $\tau \to 0$, instead of vice versa. The normal cross section of this hyperbolic cylinder is shown in Figure 6.8.

Third, consider $F_T = 0$. From Equation 6.26, this surface may be written in the form

$$\frac{RT}{\gamma - 1} = \frac{G^2\tau^2}{2} - F_x\tau + H + \frac{1}{2}\left(\frac{\tau - \tau^*}{\tau}\right)v^2 - \frac{F_y}{G}\,v \qquad (6.29)$$

The first and third portions of this surface appear exactly as the first and third parts of Figure 5.4 and need no further discussion. The function which describes the middle portion differs from the corresponding portion of Equation 5.21a in that it is divided by the variable $\tau$. When $\tau$ is large, the curvature in planes perpendicular to the $\tau$-axis now becomes constant instead of increasing linearly with $\tau$ as in the middle surface of Figure 5.4; and as $\tau$ becomes small, the middle surface now behaves

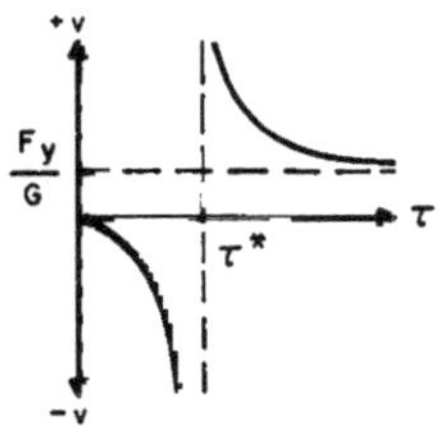


Figure 6.8. The transverse-momentum null surface ($F_v = 0$)

like a hyperbola in planes parallel to the T-τ plane.

The properties of the curve of intersection, C, between the surfaces $F_T = 0$ and $F_\tau = 0$ for the present case are very similar to the properties of the corresponding intersection between the surfaces of Equations 5.20 and 5.21, discussed in Section 3 of Chapter 5. First, the partial derivative $[R/(\gamma - 1)][\partial T/\partial\tau]_v$ from Equation 6.29 is identical to $-RT/\tau$ from Equation 6.27. Hence, the locus $(\partial T/\partial\tau)_v = 0$ of the energy null surface in the plane v-τ is exactly the intersection of the normal-momentum null surface with the plane $T = 0$. From Figure 6.7, it is clear that this locus must enclose the portion of C on which $T > 0$; therefore, everywhere on C where $T > 0$, $\partial T/\partial\tau < 0$. Thus, if the point on C where τ is a maximum lies in $T > 0$, the entire curve lies in $T > 0$, and it is always true that for variations $d\tau < 0$ on C. $dt > 0$.

As in Section 3 of Chapter 5, additional information on the shape of C is obtained from the formula for $(\partial v/\partial\tau)_C$. This is obtained by eliminating T between Equations 6.27 and 6.29 and differentiating the result with respect to τ. Then, after some reduction,

$$\left(\frac{\partial v}{\partial\tau}\right)_C = \frac{u^2 - a^2 - b_y{}^2}{-\left[\frac{(\gamma - 1)\tau}{G} F_v + \tau^* v\right]}$$

in which it is recalled from Inequalities 2.51 that $c_s{}^2 \leq a^2 + b_y{}^2$, but $c_f{}^2$ can be greater or less than $a^2 + b_y{}^2$. It is clear from the above formula that the extremum points of v on C lie at $G\tau = \sqrt{a^2 + b_y{}^2}$ On $v = 0$, it was shown in Section 3 of Chapter 5 that there are always two intersection points, say $\tau_+$ and $\tau_-$. Then, since Equation 6.25 shows that $F_v = -F_y$ when $v = 0$, it is clear that $(\partial v/\partial\tau)_C > 0$ at $\tau_+$ and $< 0$ at $\tau_-$. At the singular points, $F_v = 0$; hence, $(\partial v/\partial\tau)_C = -(u^2 - a^2 - b_y{}^2)/\tau^* v$. This shows that at $SP_1$ and $SP_4$, $(\partial v/\partial\tau)_C < 0$, and at $SP_2$ and $SP_3$ the sign of $(\partial v/\partial\tau)_C$ depends on the relative magnitudes of $\tau^*$ and $\sqrt{a^2 + b_y{}^2}/G$. From the above information, the projection of C into the v-τ plane has the form shown in Figure 6.9, in the case corresponding to case 1, Figure 5.5.

Properties of the Linear System. In this case, the linearized system is found from Equation 5.25 by eliminating the last row and column and by setting $\sigma^{-1} = 0$. The fourth equation is then solved for $\overline{B}_y^*$, which is thereupon eliminated from the remaining three. The result of these operations is

$$\begin{bmatrix} \left(u^2 - \frac{a^2}{\gamma} - b_y^{\,2} - \frac{m_1 u^2}{G}\lambda\right) & b_x b_y & \frac{a^2}{\gamma} \\ b_x b_y & \left(u^2 - b_x^{\,2} - \frac{m_2 u^2}{G}\lambda\right) & 0 \\ \frac{a^2}{\gamma} & 0 & \left(\frac{a^2}{\gamma(\gamma - 1)} - \frac{\kappa T}{G}\lambda\right) \end{bmatrix} \begin{Vmatrix} \overline{\tau}^* \\ \overline{v}^* \\ \overline{T}^* \end{Vmatrix} = 0 \tag{6.30}$$

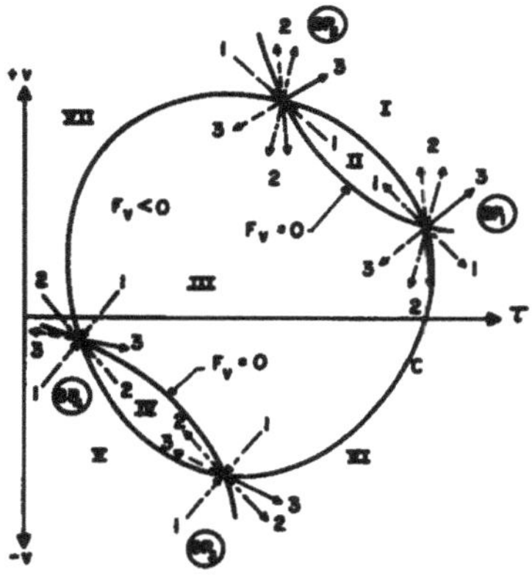


**Figure 6.9. Eigenvectors at the four singular points in the case $\sigma^{-1} = \nu = 0$**

**Since all of the diagonal elements of the above symmetric matrix contain λ's, they will all be real, and their signs can be found by applying the law of inertia of quadratic forms directly to this third-order system. As before, the quadratic form of the V-matrix (see Equation 5.26) is evaluated and manipulated into a form containing only sums of squares. The result of this now familiar procedure is**

$$V_{ij} a_i a_j = \frac{a^2}{\gamma(\gamma - 1)} \left[\overline{T}^* + (\gamma - 1)\overline{\tau}^*\right]^2 + (u^2 - b_x^{\,2})\left(\overline{v}^* + \frac{b_x b_y}{u^2 - b_x^{\,2}}\overline{\tau}^*\right)^2 + \frac{(u^2 - c_f^{\,2})(u^2 - c_s^{\,2})}{u^2 - b_x^{\,2}}\overline{\tau}^{*2} \tag{6.31}$$

from which the signs of the eigenvalues at the four singular points are

$$\left.\begin{array}{ll} SP_1: & 0 < \lambda_1 < \lambda_2 < \lambda_3 \\ SP_{2,3}: & \lambda_1 < 0 < \lambda_2 < \lambda_3 \\ SP_4: & \lambda_1 < \lambda_2 < 0 < \lambda_3 \end{array}\right\} \qquad (6.32)$$

The signs of the eigenvectors **are found from the second and** third of Equations 6.30, which give the following proportionalities:

$$\overline{\tau}^* \propto \frac{(\lambda - \lambda_{m_2})}{B_y} \overline{v}^* \propto (\lambda - \lambda_\kappa) \overline{T}^* \qquad (6.33)$$

in which

$$\left.\begin{array}{l} \lambda_{m_2} = \dfrac{G}{m_2 u^2} (u^2 \quad b_x^2) \\ \\ \lambda_\kappa = \dfrac{a^2}{\gamma(\gamma - 1)} \dfrac{G}{\kappa T} \end{array}\right\} \qquad (6.34)$$

From the equation $G\tau B_y = vB_x$, given in the first paragraph of this section, it is evident that $B_y$ and $v$ have the same sign ($B_x$ is chosen positive in all cases). Applying the discussion below Equation 5.50 to this case, we obtain

$$\left.\begin{array}{l} \lambda_3 > \max \{\lambda_{m_2}, \lambda_\kappa\} \\ \lambda_1 < \min \{\lambda_{m_2}, \lambda_\kappa\} \end{array}\right\} \qquad (6.35)$$

Finally, using 6.32, we can see that

$$\left.\begin{array}{ll} \text{at } SP_4 & \lambda_2 < \lambda_\kappa \\ \text{and at } SP_3 & \lambda_2 > \lambda_{m_2} \end{array}\right\} \qquad (6.36)$$

By the above inequalities, the relative signs of the components of $\vec{\lambda}_1$ and $\vec{\lambda}_3$ are completely determined. At $SP_4$ we know that $\overline{T}^*$ and $\overline{\tau}^*$ have opposite signs, and at $SP_3$ that $\overline{\tau}^*$ and $\overline{v}^*$ have opposite signs. Since we now know the octants in which $\vec{\lambda}_1$ and $\vec{\lambda}_3$ lie, it is possible, by use of orthogonality requirements, to locate $\vec{\lambda}_2$ at $SP_3$ and $SP_4$ completely, and at the other two singular points to locate it within one of two octants. Since Equations 6.30 are symmetrical, and there are $\lambda$'s in all diagonal elements, the $\vec{\lambda}$'s in this case are orthogonal in the ordinary sense; therefore, no two of them can lie in a single octant.

Now we can determine the relative signs of the components of each eigenvector to be as shown in Table 6.2. In the case of $\vec{\lambda}_2$

Table 6.2. Signs of the Eigenvector Components in the Case $\sigma^{-1} = \nu = 0$

| | | $\bar{T}^*$ | $\bar{\tau}^*$ | $\bar{v}^*$ |
|---|---|---|---|---|
| $SP_1$ | $\vec{\lambda}_3$ | + | + | + |
| | $\vec{\lambda}_2$ | ± | + | ∓ |
| | $\vec{\lambda}_1$ | + | - | + |
| $SP_2$ | $\vec{\lambda}_3$ | + | + | + |
| | $\vec{\lambda}_2$ | ± | + | ∓ |
| | $\vec{\lambda}_1$ | + | - | + |
| $SP_3$ | $\vec{\lambda}_3$ | + | + | - |
| | $\vec{\lambda}_2$ | + | - | + |
| | $\vec{\lambda}_1$ | + | - | - |
| $SP_4$ | $\vec{\lambda}_3$ | + | + | - |
| | $\vec{\lambda}_2$ | + | - | + |
| | $\vec{\lambda}_1$ | + | - | - |

at $SP_{1,2}$, the upper and lower signs are to be taken together, since other combinations violate the requirement of orthogonality. In the case of $\vec{\lambda}_2$ at $SP_3$, the sign of the component $\bar{T}^*$ is determined by orthogonality, and at $SP_4$ the component $\bar{v}^*$ can likewise be found.

Because the relationship between the eigenvectors and the null surfaces must be seen geometrically to carry through the existence and uniqueness proofs, we present Figure 6.9 as a summary of these data. The energy-momentum intersection curve C is shown for definiteness as that corresponding to case 1, Figure 5.5; however, the study of existence and uniqueness will not be limited to this case. We intend this figure to be viewed as a three-dimensional diagram, in which each of the regions shown contains three subregions, to be labeled with subscripts 1, 2, 3 in the order of increasing T, just as in Section 7 of Chapter 5. The solid eigenvectors are in the upper (subscript 3) region, the dot-dashed ones are in the middle regions (subscript 2), and the dotted ones are in the lower regions (subscript 1).

Behavior of the Integral Curves Away from the Singular Points. From Equations 6.24, 6.25, and 6.26, the signs of the slopes of integral curves in the various regions of τ-v-T space can be found from the following proportionalities:

$$\frac{dv}{d\tau} \propto \frac{F_v}{F_\tau}, \quad \frac{dT}{d\tau} \propto \frac{F_T}{F_\tau}, \quad \frac{dT}{dv} \propto \frac{F_T}{F_v} \tag{6.37}$$

Then, the direction in which the integral curves cross the null surfaces can be obtained by differentiating the F's with respect to x, giving

$$\left.\frac{dF_\tau}{dx}\right]_{F_\tau=0} = \frac{RG}{\tau\kappa} F_T + \frac{\tau^*}{m_2 \tau^2} vF_v \tag{6.38}$$

$$\left.\frac{dF_v}{dx}\right]_{F_v=0} = \frac{\tau^*}{m_1 \tau^2} vF_\tau \tag{6.39}$$

$$\left.\frac{dF_T}{dx}\right]_{F_T=0} = \frac{RT}{m_1 u} F_\tau - \frac{F_\tau^2}{m_1 G} - \frac{F_v^2}{m_2 G} \tag{6.40}$$

The signs of the F's can easily be found from the discussion of null surfaces. Thus, both $F_T$ and $F_\tau$ are positive in regions of higher temperature, and $F_v$ is positive when τ and v are both large or both small, and negative in the single region between.

Fast Shocks. The proof of existence and uniqueness of fast shocks can be carried through very simply in this case, in which we demonstrate that a unique transition exists between the $\vec{\lambda}_1$ at $SP_1$ and $SP_2$. Figure 6.9 shows that both of these eigenvectors lie in the region $II_2$, that $SP_1$ is an upstream node in all three eigenvector planes, and that $SP_2$ is a saddle point in the $\vec{\lambda}_1$, $\vec{\lambda}_2$ and $\vec{\lambda}_1$, $\vec{\lambda}_3$ planes. Consequently, only one integral curve can arrive at $SP_2$ in the region $II_2$. In this region, $F_T > 0$, $F_\tau < 0$, $F_v > 0$; hence, as $v > 0$ at $SP_1$ and $SP_2$, Equations 6.38, 6.39, and 6.40 show that integral curves can only leave $II_2$ through its three boundaries—they cannot enter.

But if none can enter, then the one integral curve going to $SP_2$ in the direction of the branch of $\vec{\lambda}_1$ within $II_2$ must have come from $\vec{\lambda}_1$ at $SP_1$. Thus, a shock-layer curve connecting $SP_1$ and $SP_2$ exists.

Let the integral curve which arrives at $SP_2$ along the branch of $\vec{\lambda}_1$ lying in Region $VII_2$ be denoted by C'. Then to prove uniqueness, it is only necessary to show that C' cannot come from $SP_1$. The fact that it cannot come can be seen with the help of the first of the Inequalities 6.37. In Region $VII_2$, $dv/d\tau < 0$; hence, as x decreases, C' moves upward and to the left in Figure 6.9. The slope $dv/d\tau$ cannot vanish unless C' crosses $F_V = 0$; thus C' cannot curve downward and through Region III to get to $SP_1$. If it is to cross the boundary $F_V = 0$ into Region I, however, it must first pass through a point where $d\tau = 0$. But this can happen only on $F_T = 0$, and C' cannot reach this null surface without first passing through a point $dv = 0$. Hence, it is clear that the projection of the slope of C' on the v-τ plane in $VII_2$ remains bounded away from both zero and infinity, and that it must move off to infinity in Region VII. Therefore, the shock layer found above is unique.

Slow Shocks. The configuration of eigenvectors around $SP_3$ and $SP_4$ in Figure 6.9 is identical to that of Figure 5.15. Hence, to prove existence and uniqueness of slow shocks, it is first necessary to show that the integral curves from the neighborhood around $SP_3$ actually move to the vicinity of $SP_4$, in other words, that the integral curves around these two points do not go to infinity separately without actually meshing with each other. Second, it must be proved that if the two sets of integral curves do mesh, they must do so as indicated in Figure 5.15a, and not as in 5.15b.

By drawing simple diagrams like those of Figure 5.15, we can easily show that the two families of integral curves cannot go to infinity separately unless there is a third singular point between them. Consequently, it is clear that they must mesh.

The second part of the proof is as follows: Since $dv = 0$ on $F_V = 0$, no curve lying in IV can be tangent to $F_V = 0$. This means that a family of nonmonotonic curves like those of Figure 5.15b cannot cross the null surface $F_V = 0$. Then, if the nonmonotonic curves still exist, one branch would have to bulge upward through $F_T = 0$. But Equation 6.37 shows that when both $F_T$ and $F_T$ are positive, $dT/d\tau > 0$, which is clearly impossible over the whole length of these nonmonotonic curves. Hence, Configuration 5.15b is impossible, and 5.15a is correct; as a result, a unique shock-layer curve from $SP_3$ to $SP_4$ exists. It is possible that in cases 2 and 3, Figure 5.5, the shock layer could intersect the plane $\tau = 0$; however, the general conditions under which this could happen cannot be stated in an explicit algebraic form.

Intermediate Shocks. Again, this is basically a study of the geometry of integral curves in a region including all four singular points. Compare the configuration of Figure 6.9 with that shown in Figure 6.6. If, in the present case, $\vec{\lambda}_1$ and $\vec{\lambda}_2$ at all four singular points are moved into one plane, the configuration in that plane is topologically similar to those shown in Figure 6.6. The difference in the two cases is that the $\vec{\lambda}_3$ in the present case point away from the singular points, whereas the $\vec{\lambda}_1$ in Figure 6.6 point towards them. We can easily show, by drawing a similar figure for the present case, that reversal of the normal eigenvectors produces a different pattern out of the plane of the singular points, but an entirely compatible one.

To prove that the indicated patterns of integral curves are correct in the present case, it is necessary to study the behavior of the integral curves in Regions $III_2$ and $III_3$. From Figure 6.9, it is clear that the $\vec{\lambda}_1$ at $SP_{3,4}$ lie in $III_2$, and the $\vec{\lambda}_2$ at $SP_{1,2}$ lie in either $III_2$ or $III_3$. (From Table 6.2 and Equation 6.33, we find that they lie in $III_2$ if $\lambda_{m_2} < \lambda_2 < \lambda_K$, or in $III_3$ if $\lambda_K < \lambda_2 < \lambda_{m_2}$. Now, on $F_T = 0$ in III, $F_\tau < 0$; therefore, from Equation 6.40, we see that the integral curves move from $III_2$ to $III_1$, that is, out of $III_2$. On $F_\tau = 0$ in III, $F_V < 0$ and $F_T > 0$; so, at least in the part of III where $v < 0$, the integral curves move from $III_2$ to $III_3$, that is, out of $III_2$. From these results, it is clear that the integral curves terminating in the $\vec{\lambda}_1$ at $SP_3$ and $SP_4$ in $III_2$ must have come from the part of III where $v > 0$, and also that they cannot have come up through $F_T = 0$ in any part of III. By an argument given in the preceding subsection, the integral curves from around $SP_3$ and $SP_4$ must mesh with those around $SP_1$ and $SP_2$, since there are no other singular points in between. They cannot have come from either branch of the $\vec{\lambda}_3$ at $SP_1$ and $SP_2$, because the slopes of integral curves in the regions in which the $\vec{\lambda}_3$ lie are such that integral curves from them must go to infinity. The same argument applies to all the other eigenvectors at $SP_1$ and $SP_2$ except the branches of the $\vec{\lambda}_2$ in III; hence, the curves going to the $\vec{\lambda}_1$ at $SP_3$ and $SP_4$ must have come from these branches of $\vec{\lambda}_2$.

The same existence and uniqueness properties hold for this case as in the previous one. Since $dv/d\tau$ is proportional to $m_1/m_2$ at a given location in configuration space, a small value of this ratio implies a small slope for integral curves which are remote from the null surface $F_T = 0$. In this circumstance, the integral curve topology will be that of Figure 6.6a. When $m_1/m_2$ is large, the slope $dv/d\tau$ is large and intermediate shocks will fail, as shown in Figure 6.6b. From Equation 4.38, it is clear that the minimum value of $m_1/m_2$ is 4/3; hence, in the present example, it is probable that the more common configuration will be one in which there are no intermediate shocks.

4. Negligible Bulk Viscosity and Shear Viscosity

The equations of this special case are obtained from Equations 5.13 - 5.17 by setting $m_1 = m_2 = 0$. Then $v$ can be eliminated by means of Equation 5.14, and Equation 5.13 can be used to simplify Equation 5.17. (It could also be used to modify Equation 5.15, but there is no advantage in doing so.) Hence, the basic equations of the present problem are

$$0 = \frac{RT}{\tau} + G^2\tau + \frac{B_y^2}{2\mu} - F_x = F_\tau \tag{6.41}$$

$$\frac{\kappa}{G}\frac{dT}{dx} = \frac{RT}{\gamma - 1} - \frac{G^2\tau^2}{2} + F_x\tau - H - \frac{\tau B_y^2}{2\mu} + \frac{1}{2G^2}\left(\frac{B_x}{\mu}B_y + F_y\right)^2 = F_T \tag{6.42}$$

$$\frac{1}{\mu}\frac{dB_y}{dx} = J_z \tag{6.43}$$

$$\nu G\tau^2\frac{dJ_z}{dx} = -\frac{J_z}{\sigma} + GB_y(\tau - \tau^*) - \frac{F_yB_x}{G} = F_J \tag{6.44}$$

Null Surfaces. In this problem, the primary variables are $T$, $B_y$, and $J_z$; and there is a secondary variable $\tau$. Although the specific volume $\tau$ could be eliminated by solving the quadratic 6.41 for its one positive root, this is not done because the result would be a very unwieldy set of null surfaces in the three-space $T$-$B_y$-$J_z$. The relationship of this example with the others is much easier to see if this problem is considered in a four-space of the variables $T$, $B_y$, $J_z$, and $\tau$. Then, to obtain a partial visualization of the family of null surfaces, they can be projected onto the simple null surface $J_z = 0$, just as explained in Chapter 5. In fact, the entire discussion of null surfaces of Chapter 5 is applicable here, the only differences in this case being: (1) that just one projection — from a four-space to a three-space — need be made to visualize the system, and (2) that the integral curves must either lie on $F_\tau = 0$ or be parallel to the $\tau$-axis. The latter behavior indicates that there may be a discontinuity in $\tau$, that is, a subshock. Suppression of the coordinate $J_z$ is not serious, because the qualitative behavior of $J_z(x)$ is easily seen by noting that $J_z$ vanishes if and only if an integral curve passes through a point where $B_y$ is stationary (that is, $dB_y = 0$), and $dJ_z/dx$ vanishes if and only if an integral curve crosses the null surface $F_J = 0$.

Properties of the Linearized System. For this case, the system linearized about a singular point is obtained from Equation 5.25

by setting $m_1 = m_2 = 0$. After elimination of the secondary eigenvector components $\overline{\tau}^*$ and $\overline{v}^*$, those equations become

$$\begin{bmatrix} \frac{\kappa T}{G}(\lambda_\kappa - \lambda) & -\frac{a^2}{\gamma}\frac{b_x b_y}{\left(u^2 - \frac{a^2}{\gamma}\right)} & 0 \\ -\frac{a^2}{\gamma}\frac{b_x b_y}{\left(u^2 - \frac{a^2}{\gamma}\right)} & \frac{b_x^2}{\mu\sigma u}(\lambda_\sigma - \lambda) & -\frac{m_-}{e}\frac{u^2}{G}\frac{B_x}{\mu}\lambda \\ 0 & -\frac{m_-}{e}\frac{u^2}{G}\frac{B_x}{\mu}\lambda & \frac{m_-}{m_+}u^2 \end{bmatrix} \begin{Vmatrix} \overline{T}^* \\ \overline{B}_y^* \\ \overline{J}_z^* \end{Vmatrix} = 0 \qquad (6.45)$$

in which

$$\lambda_\kappa = \frac{a^2}{\gamma(\gamma - 1)}\left(\frac{u^2 - a^2}{u^2 - \frac{a^2}{\gamma}}\right)\frac{G}{\kappa T} \qquad (6.46)$$

$$\lambda_\sigma = \left(\frac{u^2 - b_x^2}{u^2} - \frac{b_y^2}{u^2 - \frac{a^2}{\gamma}}\right)\mu\sigma u \qquad (6.47)$$

It will also be useful to have the equation

$$\overline{\tau}^* = \frac{\gamma\kappa T}{a^2 G}\left[\lambda - \frac{a^2 G}{(\gamma - 1)\gamma\kappa T}\right]\overline{T}^* \qquad (6.48)$$

The factor $a^2G/(\gamma - 1)\gamma\kappa T$ in this equation is larger than $\lambda_\kappa$ if $u^2 > a^2/\gamma$, and less in the opposite case.

As usual, $D_2(\lambda)$ is defined as the determinant of the first two rows and columns of the matrix in 6.45. Then, when the determinant of Equations 6.45 is equated to zero, it may be written

$$D_2(\lambda) + c\lambda^2(\lambda - \lambda_\kappa) = 0 \qquad (6.49)$$

where $c$ is a positive constant.

To find the nature of the eigenvalues of Equation 6.49, the signs of the real eigenvalues of $D_2(\lambda) = 0$ must first be determined. Again, this can be done most easily by applying the law of inertia of quadratic forms. The appropriate quadratic form in Equations 6.45 can be manipulated into the following sum of squares:

$$V_{ij}a_i a_j = \frac{a^2}{\gamma(\gamma - 1)}\left(\frac{u^2 - a^2}{u^2 - \frac{a^2}{\gamma}}\right)\left[\overline{T}^* - \frac{(\gamma - 1)\,b_x b_y}{u^2 - a^2}\,\overline{B}^*\right]^2$$

$$+ \frac{b_x^{\,2}(u^2 - c_f^{\,2})(u^2 - c_s^{\,2})}{u^2(u^2 - a^2)}\,\overline{B}_y^{*\,2} \qquad (6.50)$$

The behavior of the signs of $D_2(\lambda) = 0$ is clearly much more complicated than in the previous example, since they now depend on the isentropic sound speed $a$ and the isothermal sound speed $a/\sqrt{\gamma}$. Also, an additional complication arises from the fact that, while $a$ is always between $c_f$ and $c_s$, $a/\sqrt{\gamma}$ may be less than $c_s$. The easiest way to display the signs of the eigenvalues of $D_2(\lambda) = 0$ seems to be by the diagram in Figure 6.10. The plus and minus signs indicate the signs of the two eigenvalues in each of six regions, lettered A, B, C, D, E, F. The singular points which can lie in each region are also indicated.

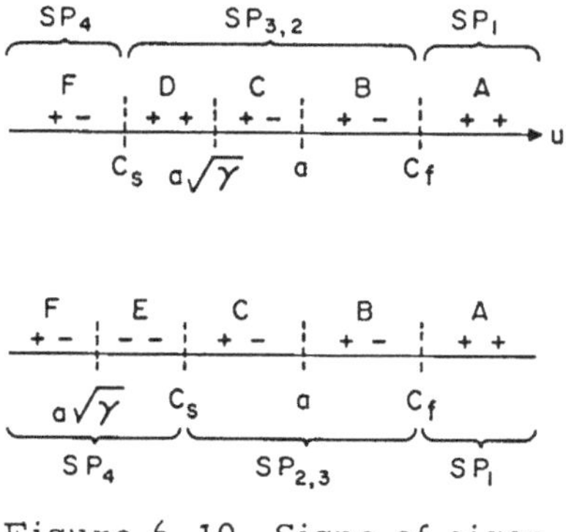


Figure 6.10. Signs of eigenvalues of $D_2(\lambda) = 0$

Now that the signs of $D_2(\lambda) = 0$ are known, the usual graphical procedure can be used to find the nature of the roots of Equation 6.49. Only the results are given, because it will be more profitable for the interested reader to verify them himself by drawing the required simple graphs. Each of four possible locations of the origin ($\lambda = 0$) and of $\lambda_\kappa$ relative to the three eigenvalues of Equation 6.45 is given below. The identifying regions of Figure 6.10 and corresponding singular points are listed alongside each set of inequalities:

$$\left.\begin{array}{ll} A|SP_1\ ;\ D|SP_{2,3} & \lambda_1 < 0 < \lambda_2 < \lambda_K < \lambda_3 \\ B|SP_{2,3}\ ;\ F|SP_4 & \lambda_1 < \lambda_2 < 0 < \lambda_K < \lambda_3 \\ C|SP_{2,3} & \lambda_1 < \lambda_2 < \lambda_K < 0 < \lambda_3 \\ E|SP_4 & \lambda_1 < \lambda_2 < \lambda_K < \lambda_3 < 0 \end{array}\right\} \quad (6.51)$$

In the cases here for which both $\lambda_1$ and $\lambda_2$ are negative, they can also be complex with negative real parts.

The relative signs of the eigenvector components can be found from the first and third of Equations 6.45. Also, from Equation 6.48 the sign of $\bar{\tau}^*$ relative to $\bar{T}^*$ can sometimes be found. Then, making use of Figure 6.10 and the results of Inequalities 6.51, it can easily be verified that the signs of the eigenvectors are the same as or opposite those indicated in Table 6.3. The dots in this table indicate cases in which there is not enough information available to determine the sign of $\bar{\tau}^*$

Table 6.3. Signs of Eigenvector Components for the Special Case in Which $m_1 = m_2 = 0$

| Singular Point | Region (Figure 6.10) | $\vec{\lambda}$ | $\bar{\tau}^*$ | $\bar{T}^*$ | $\bar{B}_y^*$ | $\bar{J}_z^*$ |
|---|---|---|---|---|---|---|
| $SP_1$ | A | 3 | · | + | - | - |
| | | 2 | - | + | + | + |
| | | 1 | - | + | + | - |
| $SP_2$ | B, C | 3 | · | + | - | - |
| | | 2 | - | + | + | - |
| | | 1 | - | + | + | - |
| | D | 3 | + | + | + | + |
| | | 2 | · | + | - | - |
| | | 1 | - | + | - | + |
| $SP_3$ | B, C | 3 | · | + | + | + |
| | | 2 | - | + | - | + |
| | | 1 | - | + | - | + |
| | D | 3 | + | + | - | - |
| | | 2 | · | + | + | + |
| | | 1 | - | + | + | - |
| $SP_4$ | E | 3 | - | + | + | - |
| | | 2 | - | + | - | + |
| | | 1 | - | + | - | + |
| | F | 3 | + | + | - | - |
| | | 2 | - | + | + | - |
| | | 1 | - | + | + | - |

It will be convenient, just as in the previous example, to have the locations of the eigenvectors at each of the singular points summarized graphically. Because two sets of eigenvectors are possible at each of the three lower singular points, the graphical summary will obviously be more complicated. Figure 6.10 shows that there are four cases to consider, depending on whether $a/\sqrt{\gamma}$ is below $SP_4$, between $SP_4$ and $SP_3$, between $SP_3$ and $SP_2$, or between $SP_2$ and $SP_1$. In Figure 6.11, the first and last of these cases are shown, and from them the two intermediate cases can be inferred. In this example all of the eigenvectors lie on the surface $F_\tau = 0$; hence, there is no ambiguity in showing them all as solid arrows. The direction of the $J_z$-component can be inferred from the fact that $J_z \propto dB_y$.

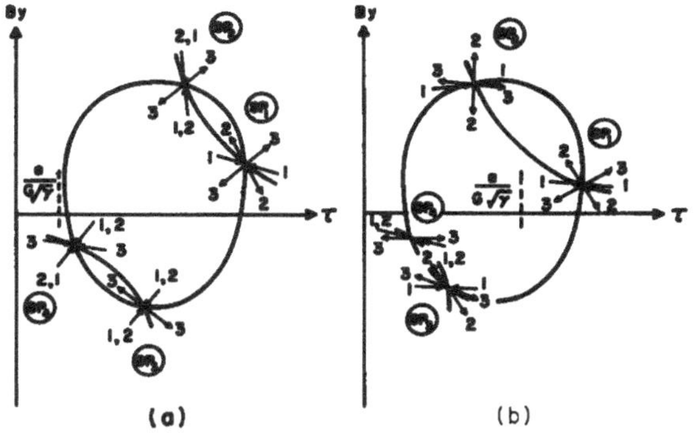

Figure 6.11. Eigenvectors at the four singular points in the case $m_1 = m_2 = 0$

**<u>General Properties of the Integral Curves.</u> The behavior of the integral curves away from the singular points can be found, as usual, by considering the signs of their slopes in the various regions defined by the null surfaces, and by evaluating the total derivatives of the F-functions on the null surfaces. In this case some special considerations also arise.**

**The slope $d\tau/dB_y$ is found directly by dividing Equation 6.42 by Equation 6.43, which shows that**

$$\frac{d\tau}{dB_y} \propto \frac{F_\tau}{J_z} \tag{6.52}$$

To obtain the other two slopes, Equation 6.41 is first differentiated with respect to x. This gives

$$\frac{d\tau}{dx} = - \tau^2 \frac{N}{D} \tag{6.53}$$

in which

$$\left.\begin{aligned} N &= B_y J_z + \frac{RG}{\tau\kappa} F_T \\ D &= u^2 - \frac{a^2}{\gamma} = G^2\tau^2 - RT \end{aligned}\right\} \qquad (6.54)$$

Then, if Equation 6.43 is divided by Equation 6.53,

$$\frac{dB_y}{d\tau} = - \frac{\mu J_z}{\tau^2} \frac{D}{N} \qquad (6.55)$$

Finally, from Equations 6.52 and 6.54,

$$\frac{dT}{d\tau} \propto - F_T \frac{D}{N} \qquad (6.56)$$

We can calculate the total derivative of $F_T$ with respect to $x$ at $F_T = 0$ from Equation 6.42; and after taking Equation 6.41 into account, we can manipulate it into the form

$$\left.\frac{dF_T}{dx}\right]_{F_\tau = F_T = 0} = - J_z \left[ \frac{\frac{a^2}{\gamma}}{u^2 - \frac{a^2}{\gamma}} \tau B_y + \frac{1}{G} (F_J)_{J_z = 0} \right] \qquad (6.57)$$

in which the definition of the second term is found from Equation 6.44. Equating $(F_J)_{J_z=0}$ to zero, we obtain the pair of hyperbolas of Figure 5.11, between which the function $(F_J)_{J_z=0}$ is negative.

Important information on the behavior of the integral curves can be obtained directly from Equation 6.53. A type of singular curve occurs as the intersection of the two surfaces $N = 0$ and $D = 0$, because the slope $d\tau/dx$ is undefined there; however, when integral curves cross $D = 0$ at points where $N \neq 0$, the slope $d\tau/dx$ becomes infinite and then changes sign, indicating an extremum in $x$. But if an integral curve is to correspond to a shock layer, $x$ must be monotone along it; hence, none of the curves for which there is a point $D = 0$, $N \neq 0$ can correspond to shock layers. For the sheet of integral curves which simultaneously passes through both $D = 0$ and $N = 0$, however, the sign of $d\tau/dx$ is preserved, and there is no reason why one of such curves could not correspond to a shock layer. It is also evident that the surface $N = 0$ is a locus of extremum points for

$\tau$, and that it coincides with $F_T = 0$ when either $J_z$ or $B_y$ is zero. The surface $D = 0$ is a parabolic cylinder with generators parallel to the $B_y$-axis in the usual $T$-$\tau$-$B_y$ space.

Existence and Uniqueness of the Shock Layers. It will be more convenient in this case, in view of the results of Sections 2 and 3 of this chapter, to discuss all types of transitions simultaneously. The procedure will be first to consider the topological configuration of eigenvectors and to compare them with previous configurations. Then, after obtaining the mesh of integral curves which satisfies all requirements in the large, we shall analyze and discuss properties of the resulting shock-layer curves in the following subsection.

First, consider the case shown in Figure 6.11a. At $SP_1$, there are two positive eigenvalues and one negative eigenvalue; at $SP_2$ and $SP_3$, there are one positive eigenvalue and two negative eigenvalues; and at $SP_4$ all of the eigenvalues are negative. Comparison with Figure 6.6 makes it clear that this is exactly the situation in the example of Section 2, and one anticipates that the same transitions will exist. As further evidence, it can be seen from Figure 6.11 that the $\vec{\lambda}_2$ at $SP_1$ and $SP_2$ are positioned correctly to yield a fast-shock transition, the $\vec{\lambda}_3$ at $SP_3$ and $SP_4$ are positioned to yield a slow-shock transition, the $\vec{\lambda}_3$ at $SP_1$ and $SP_2$ are positioned to initiate intermediate shocks, and the $\vec{\lambda}_2$ at $SP_3$ and $SP_4$ to terminate intermediate shocks. The integral curves in each of the regions around the singular points are monotone in this case (the singular curve at $u = a/\sqrt{\gamma}$ is off to the left); hence, it is not difficult to show, by drawing all conceivable configurations of integral curves, that the configurations of Figure 6.6 are the only possible ones. Therefore, the existence and uniqueness properties of Figure 6.6 apply to the case of Figure 6.11a.

Second, consider the case shown in Figure 6.11b. At $SP_2$, $SP_3$, and $SP_4$, the signs of the eigenvalues are exactly those of the example of Section 3 (see Equation 6.32), but at $SP_1$ the signs are still those of Section 2. This case is shown in normal form in Figure 6.12 for the case in which there are intermediate shocks. In the plane of the singular points, the compatible configurations are still the usual ones, and hence the same existence and uniqueness properties apply. The integral curves between

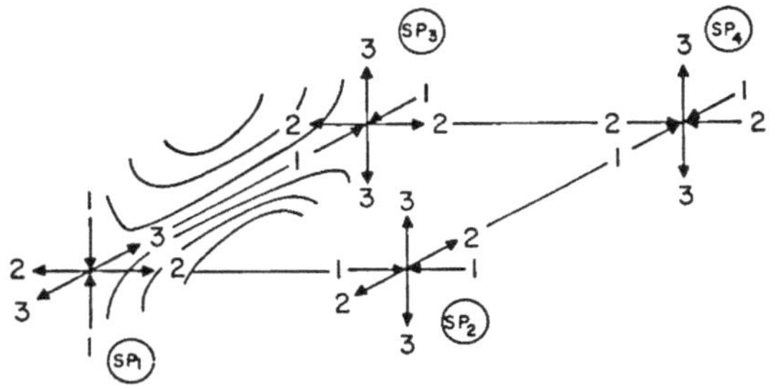


Figure 6.12. Eigenvector configuration in the case of Figure 6.11b

$SP_1$ and the other singular points lying out of the plane of the four singular points must be as shown in Figure 6.12, in which extremum points in $B_y$ and $T$ (see Equations 6.55 and 6.56) occur on the surface $u^2 = a^2/\gamma$ whenever $N \neq 0$ (see Equation 6.54). Only by such a configuration can the integral curves mesh in a compatible way.

The configurations of integral curves in the cases in which the surface $u^2 = a^2/\gamma$ lies between $SP_2$ and $SP_3$ or between $SP_3$ and $SP_4$ are similar to that shown in Figure 6.12, i.e., at the singular points to the high-$\tau$ side of $u^2 = a^2/\gamma$, the eigenvector configurations are those of Section 2, and to the low-$\tau$ side they are those of Section 3. Extremum points in $B_y$ and $T$ occur on $u^2 = a^2/\gamma$, and the configurations of integral curves in the plane of the singular points are the same. Thus, unique $1 \rightarrow 2$ and $3 \rightarrow 4$ shock layers always exist; unique $1 \rightarrow 3$ and $2 \rightarrow 4$ shocks and an infinite family of $1 \rightarrow 4$ shocks exist when $\kappa$ is small compared with $\nu$ and $\sigma^{-1}$, and a unique $2 \rightarrow 3$ shock exists in the boundary case between the above-described intermediate shocks and the case in which there are no intermediate shocks.

Properties of the Shock Layers. In this section, the qualitative properties of each type of shock will be discussed. First, let us examine the two types of fast shocks. In Figure 6.11a, the unique transition goes between the $\vec{\lambda}_2$ and will be monotone in $T$, $\tau$, and $B_y$, with $J_z > 0$ throughout, if $\lambda_2$ is real, or there will be downstream spatial oscillations in the flow if $\lambda_2$ is complex. (These conclusions are easy to check by means of procedures that are well established and that have been repeated several times already.) For fast shocks in Figure 6.11b, the transition goes from $\vec{\lambda}_2$ at $SP_1$ to $\vec{\lambda}_1$ at $SP_2$. The positive slope of $\vec{\lambda}_1$ at $SP_2$ indicates that $J_z < 0$ there: hence, there is a point along the transition at which $J_z = 0$, after which $dB_y/d\tau > 0$. But in the full five-dimensional problem, $dB_y/d\tau < 0$ throughout the transition for all finite values of the dissipation coefficients. Consequently, the fast-shock layer in Figure 6.11b cannot be attained in a continuous manner as $m_1$, $m_2$ approach zero. This implies the existence of a subshock, i.e., when $B_y$ reaches its value at $SP_2$, the shock-layer curve goes directly to $SP_2$ along a line $dB_y = dT = dJ_z = 0$, $d\tau \neq 0$.

Second, let us consider the three types of slow shocks. In Figure 6.11a, the transition goes between the $\vec{\lambda}_3$ and is always monotone in $T$, $\tau$, and $B_y$. In Figure 6.11b, the transition goes between the $\vec{\lambda}_2$ and may be either monotone or oscillatory depending on whether $\lambda_2$ at $SP_4$ is real or complex. In the third type of slow shock, the surface $u^2 = a^2/\gamma$ separates $SP_3$ and $SP_4$; however, the properties of the shock are still the same as for the second type.

Finally, consider intermediate shocks. In Figure 6.11a, the $1 \rightarrow 3$ and $2 \rightarrow 4$ shocks are monotone (unless the $\lambda_2$ at $SP_3$ and

$SP_4$ are complex), and the $1 \rightarrow 4$ shocks undergo a sign reversal in $J_z$ just as in the five-dimensional case. In Figure 6.11b, all of the intermediate shocks undergo sign reversals in $J_z$. For the other cases, the properties can be similarly deduced.

## 5. Magnetic Field Parallel to Plane of Shock Wave

The Shock-Layer Equations. The present case is characterized by the condition $B_x = 0$. When this is true, it is no longer possible to choose a reference frame in which the tangential components of the electric field are zero, but, on the other hand, it is possible to translate the coordinate system parallel to the shock at a velocity such that $F_y = F_z = 0$ (see Equations 4.32, 4.33). This coordinate system can be rotated about the x-axis until $E_y = 0$; then, with $F_y = E_y = B_x = 0$, the only bounded solution of Equation 4.32 is $v = 0$. Thus, the shock-layer equations for the present case can be found from the equations in Section 2 of Chapter 4 by setting $F_y = E_y = B_x = v = 0$. Then, by the same arguments used in the general case, the $E_x$, $J_x$, and $w$ coupling terms can be dropped for existence studies. With these simplifications, four equations remain for the qualitative study of shocks: Equations 4.31, 4.34, 4.43, and 4.49. Using the auxiliary equations 4.30, 4.35, and 4.39, and making the above-indicated deletions, the shock-layer equations for this case are

$$m_1 G \frac{d\tau}{dx} = \frac{RT}{\tau} + G^2\tau + \frac{B_y^2}{2\mu} - F_x = F_\tau \tag{6.58}$$

$$\frac{\kappa}{G}\frac{dT}{dx} = \frac{RT}{\gamma - 1} - \frac{G^2\tau^2}{2} + F_x\tau - H - \frac{\tau B_y^2}{2\mu} - \frac{E_z}{G\mu} B_y = F_T \tag{6.59}$$

$$\frac{1}{\mu}\frac{dB_y}{dx} = J_z \tag{6.60}$$

$$\nu G\tau^2 \frac{dJ_z}{dx} = -\frac{\nu\tau}{m_1} J_z F_\tau - \frac{J_z}{\sigma} + G\tau B_y + E_z = F_J \tag{6.61}$$

Null Surfaces. Equations 6.58 and 6.60 are identical with Equations 5.13 and 5.16, respectively, and need no further discussion. Comparing the equation $F_T = 0$, from Equation 6.59, with Equation 5.21a, we can see that the forms of the two would be identical if $\tau^* = 0$ in Equation 5.21a, and if we choose $E_z$ to be negative. Hence, we can easily visualize the form of the energy null surface from Figure 5.4.

Consider the null surface of Equation 6.61: $F_J = 0$. The

projection of this surface into the three-space $J_z = 0$ is a hyperbolic cylinder lying in $\tau > 0$, $B_y > 0$ (if $E_z < 0$), with generators parallel to the T-axis. Away from the space $J_z = 0$, the form of $F_J = 0$ can be visualized in a $B_y$-$\tau$-$J_z$ space defined by the condition $F_T = 0$, in exactly the manner described in Section 2. With this picture and the fact of continuity, this null surface is known well enough for the existence proofs.

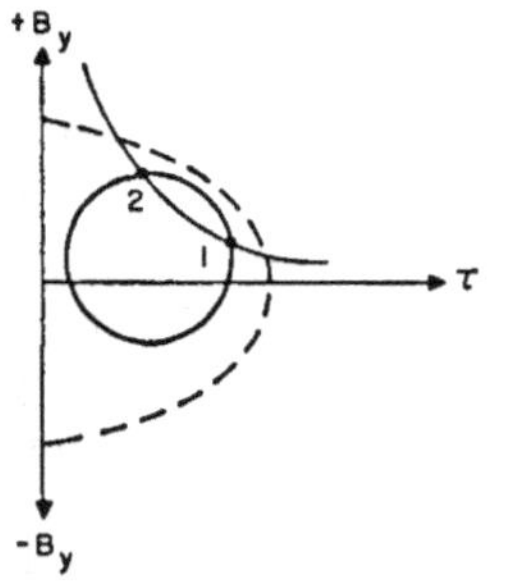


Figure 6.13. Intersection of the null surfaces in the case $B_x = 0$

It is clear from the shapes of the three null surfaces that there are only two singular points, both lying in the space $J_z = 0$. When both singular points are present, the projections onto $J_z = 0$ of the intersections of the null surfaces appear as the solid curves shown in Figure 6.13, in which the T-axis extends upwards from the paper. The dotted curve is the intersection of $F_T = 0$ in the plane $T = 0$, and is shown for reference.

Properties of the Linear System. Consider the dimensionless variables defined by Equations 5.24. The first, third, and fifth will be retained for this problem, but, since $B_x = 0$ now, the fourth will be replaced by

$$B_y = B_{y_i}(1 + B_y^*) \tag{6.62}$$

The form of the fourth-order linear system for the present case can then be obtained directly from Equation 5.25 by comparing Equations 5.13 - 5.17 with the present set of shock-layer equations.

First, since there is no longer an equation for v, we delete the second row and column of Equation 5.25. Then, since Equations 5.13 and 6.58 are identical, the only difference in the linearized form of this equation is due to the new definition in Equation 6.62, which shows that the magnetic term $B_{y_i}^2$ is now inserted in place of $B_xB_{y_i}$. Next, all terms containing $B_y$ cancel out of both energy equations when the system is linearized; hence, the third row of Equation 5.25 remains unchanged. Then, because Equations 5.16 and 6.60 are identical, the only change in the last of Equations 5.25 is to replace $B_x$ by $B_{y_i}$. Finally, compare Equations 5.17 and 6.61. The constant $E_z$ does not appear in the linearized system, and v = 0 now; thus, when we remember that the second column of Equations 5.25 has been removed, the last term in Equations 5.17 and 6.61 contributes no difference. The only difference is in the term $G\tau B_y$, in which the linearized term proportional to $\overline{B}_y^*$ is now multiplied by a constant $B_{y_i}$ instead of $B_x$. Also, the fourth of Equations 5.25 was multiplied by the factor $B_x$, which is now replaced by $B_{y_i}$. Making these changes, we can see that the linear system for this problem can be obtained from Equations 5.25 by deleting the second row and column and by replacing $B_x$ by $B_{y_i}$ wherever it appears; hence,

$$\begin{bmatrix} \left(u^2 - \frac{a^2}{\gamma} - \frac{m_1 u^2}{G}\lambda\right) & \frac{a^2}{\gamma} & b_y^{\ 2} & 0 \\ \frac{a^2}{\gamma} & \left(\frac{a^2}{\gamma(\gamma-1)} - \frac{\kappa T}{G}\lambda\right) & 0 & 0 \\ b_y^{\ 2} & 0 & b_y^{\ 2}\left(1 - \frac{\lambda}{\mu\sigma u}\right) - \left(\frac{m_-}{e}\frac{u^2}{G}\frac{B_y}{\mu}\right)\lambda & -\left(\frac{m_-}{e}\frac{u^2}{G}\frac{B_y}{\mu}\right)\lambda \\ 0 & 0 & -\left(\frac{m_-}{e}\frac{u^2}{G}\frac{B_y}{\mu}\right)\lambda & \frac{m_-}{m_+}u^2 \end{bmatrix} \begin{Vmatrix} \overline{\tau}^* \\ \overline{T}^* \\ \overline{B}_y^* \\ \overline{J}_z^* \end{Vmatrix} = 0 \tag{6.63}$$

To determine the nature of the eigenvalues, we follow the usual procedure. First, we define $D_3(D_2)$ as the determinant of the first three (two) rows and columns of Equations 6.63. Then, equating the determinant of Equations 6.63 to zero, we obtain

$$D_3(\lambda) - c\lambda^2 D_2(\lambda) = 0 \tag{6.64}$$

in which $c > 0$, and we note that $D_3(\lambda) < 0$ for large $\lambda$.

The signs of $D_3(\lambda) = 0$ can be found from the quadratic form

$$V_{ij}a_i a_j = \frac{a^2}{\gamma(\gamma - 1)}\left[\overline{T}^* + (\gamma - 1)\overline{\tau}^*\right]^2 + b_y^{\ 2}(\overline{B}_y^* + \overline{\tau}^*)^2 + (u^2 - a^2 - b_y^{\ 2})\overline{\tau}^{*2} \tag{6.65}$$

But $\sqrt{a^2 + b_y^{\ 2}}$ is the speed of magnetoacoustic waves in the direction perpendicular to the field lines, i.e., in the x-direction; and for finite amplitude shocks, the upstream velocity is greater than $\sqrt{a^2 + b_y^{\ 2}}$ and the downstream velocity less. Consequently, $u^2 > a^2 + b_y^{\ 2}$ at $SP_1$, and $u^2 < a^2 + b_y^{\ 2}$ at $SP_2$. Thus, at $SP_1$ all three real eigenvalues of $D_3(\lambda) = 0$ are positive, and at $SP_2$ two are positive and one is negative.

The signs of $D_2(\lambda) = 0$ can be found from a quadratic form obtained from Equation 6.65 by setting $\overline{B}_y^* = 0$. The result shows that both signs are positive when $u > a$, and one is positive and one negative if $u < a$.

With this information, the signs of the eigenvalues of Equation 6.64 can be obtained by the usual graphical procedure. They are as follows:

$$\left.\begin{aligned} SP_1: &\quad \lambda_1 < 0 < \lambda_2 < \lambda_3 < \lambda_4 \\ SP_2: &\quad \lambda_1 < \lambda_2 < 0 < \lambda_3 < \lambda_4 \end{aligned}\right\} \qquad (6.66)$$

At $SP_1$, all of the $\lambda$'s are real; and at $SP_2$, $\lambda_1$ and $\lambda_2$ may be either real or complex, always with negative real parts.

The relative signs of the components of the eigenvectors can be found from the first, second, and fourth of Equations 6.63, which show that

$$\left.\begin{aligned} \overline{B}_y^* &\propto D_2(\lambda)\, \overline{T}^* \\ \overline{\tau}^* &\propto (\lambda - \lambda_K)\, \overline{T}^* \\ \overline{J}^* &\propto \lambda\, \overline{B}_y^* \end{aligned}\right\} \qquad (6.67)$$

The definition of $\lambda_K$ is obvious from the second of Equations 6.63. From the graphical solution for the signs of the eigenvalues, it is clear that $D_2(\lambda)$ is positive at $\lambda_1$, $\lambda_2$, $\lambda_4$ and negative at $\lambda_3$. The quantity $\lambda - \lambda_K$ is positive at $\lambda_4$, negative at $\lambda_1$, $\lambda_2$, and indefinite at $\lambda_3$. With this information, Table 6.4 for the relative signs of the eigenvector components, can be constructed.

Table 6.4 Relative Signs of Eigenvector Components when $B_x = 0$

| | | $\overline{\tau}^*$ | $\overline{T}^*$ | $\overline{B}_y^*$ | $\overline{J}_z^*$ |
|---|---|---|---|---|---|
| $\vec{\lambda}_1$ | $SP_{1,2}$ | - | + | + | - |
| $\vec{\lambda}_2$ | $SP_1$ | - | + | + | + |
| | $SP_2$ | - | + | + | - |
| $\vec{\lambda}_3$ | $SP_{1,2}$ | · | + | - | - |
| $\vec{\lambda}_4$ | $SP_{1,2}$ | + | + | + | + |

Existence and Uniqueness of the Shock Layer. The study of the behavior of integral curves in the large follows along exactly the same lines as for the general case of Chapter 5, the only difference being that the study is now simpler because there is one less dimension. The formula from Equation 6.58 for the total derivative of $F_\tau$ on $F_\tau = 0$ is exactly that of Equation 5.84, and the corresponding formula from $F_T$ is obtained from Equation 5.86 by setting $F_v = 0$.

The proof of existence and uniqueness for the present case differs from the corresponding proof in Section 8 of Chapter 5 for fast shocks only in that there is one less positive eigenvector at each of the singular points $SP_1$ and $SP_2$. Other than that, the details of the proof are the same and will not be repeated here. **The result is that a unique shock layer exists.**

## 6. "Switch-on" and "Switch-off" Shocks

**This important special case is characterized by the condition** $F_y = 0$. From the shock-layer equations 5.13 through 5.17, it is clear that only the transverse-momentum equation 5.14, and the energy equation 5.15 are affected by this condition. At the singular points, Equations 5.14 and 5.17 now become

$$\left.\begin{aligned} Gv &= \frac{B_x}{\mu} B_y \\ B_x v &= G\tau B_y \end{aligned}\right\} \qquad (6.68)$$

which show that either $v = B_y = 0$ or $u^2 = B_x^{\,2}\,\tau/\mu$. Thus, as mentioned in Chapter 2, these shocks are characterized by the fact that the flow on one side is normal, and on the other side oblique. On the side on which it is oblique, $u$ is equal to the normal Alfvén speed.

Consider the null surfaces projected into the three-space $T$-$\tau$-$B_y$ in which $F_v = J_z = 0$. The projected $F_J$ surface, Equation 5.22, then reduces to the two planes $B_y = 0$ and $\tau = \tau^*$, consistent with the above results. In the $T$-$\tau$-$B_y$ three-space, this appears to be a profound change in $F_J = 0$; however, the five-space null surfaces obtained from Equations 5.13-5.17 show no such drastic behavior. The $F_J$ null surface is, in reality, not affected at all; it is merely projected on a particular plane ($F_v = 0$) which gives it an unusual appearance. From Equation 5.21a, it is evident that the energy null surface $F_T = 0$ is now simplified in that it has become symmetrical.

Since $F_y$ does not appear in Equations 5.25, its value has no direct effect on the eigenvalues and eigenvectors. The only implicit effect upon them is due to the fact that the singular points have moved. In addition, $F_y$ does not appear in Equations 5.84

through 5.88, nor has it any effect on the signs of the slopes of the integral curves in the various regions defined by the null surfaces. Because of these considerations and the fact that there is nothing "singular" about the limit $F_y \to 0$, it is clear that the corresponding integral curves must be continuous variations from the case $F_y \neq 0$. Hence, all transitions which exist when $F_y \neq 0$ also exist when $F_y = 0$.

## 7. Interpretation of the Results

We indicated at the beginning of this chapter that the purpose in giving the examples of Sections 2, 3, and 4 was to try to shed more light on existence and uniqueness properties of shock layers in the full five-dimensional problem. In this section, we shall review the similarities and differences between these examples, keeping in mind that each is a different type of limit of the five-dimensional problem.

In the plane of the singular points (topologically speaking), the behavior in all three examples is similar. Unique fast and slow shock layers exist in all of these examples; and in cases 2 and 3 of Figure 5.5, the slow shock-layer curve may sometimes intersect the plane $\tau = 0$, resulting this way in nonexistence of the shock. When $m_1$ and $\kappa$ are small compared with the other dissipation coefficients, intermediate shock layers exist as well; but in the opposite case they fail to exist. In all of the examples there is a boundary case for which a unique $2 \to 3$ shock layer exists. The differences outside the plane of the singular points are due to the directions of the eigenvectors perpendicular to the plane of the square. In Section 2 these perpendicular eigenvectors all point toward their respective singular points, and in Section 3 they all point away. In Section 4 those at singular points below the surface $G\tau = a/\sqrt{\gamma}$ point away from the singular points, and those above point toward them. (The reason that this mixed configuration was compatible with all of the topological requirements for integral curves, discussed in Section 9 of Chapter 5, was the existence of the auxiliary singular surfaces defined as the zeros of the numerator and denominator of Equation 6.53.)

With respect to all pairs of the coefficients $m_1$, $m_2$, $\kappa$, $\sigma^{-1}$, $\nu$, it made physical sense only to set the three pairs $m_2$, $\kappa$; $\sigma^{-1}$, $\nu$; $m_1$, $m_2$ equal to zero, while the other three were greater than zero. In all of these physically possible limit cases (the examples of Sections 2, 3, and 4), the integral-curve patterns in the plane of the singular points were topologically identical, the result being that the existence and uniqueness properties of shocks in these three limit cases were identical. A little more insight into the reasons for the differences outside the plane of the singular points can be seen by considering what happens to the eigenvalues of the fifth-order system (Equations 5.25) as the various pairs of coefficients vanish. First, when $m_2$, $\kappa \to 0$, the factors $G/m_2$ and $Ga^2/\gamma(\gamma - 1)\kappa T$ in the second and third of Equations 5.25 go to $+\infty$. But from the theory of eigenvalues, discussed in Chapter 5, the largest eigenvalue of the fifth-order system must always be greater than both of the above factors. Hence, suppose $m_2 \to 0$ first; then, $\lambda_5 \to +\infty$ and a fourth-order system remains. When $\kappa \to 0$, $\lambda_4 \to +\infty$ and a third-order system remains; the signs of the eigenvalues of this system must be those corresponding to the three lowest eigenvalues of the fifth-order system. Second, consider the case $\sigma^{-1}$, $\nu \to 0$. When $\nu \to 0$, $\lambda_1 \to -\infty$, as has been shown by the graphical procedure of Chapter 5. Then, when $\sigma^{-1} \to 0$, $\lambda_5 \to +\infty$; thus the signs of the eigenvalues of the remaining third-order system are those of the three intermediate eigenvalues of the fifth-order system. Finally, when $m_1$, $m_2 \to 0$, the first of Equations 5.25 shows that an eigenvalue moves to $+\infty$ if $u^2 > a^2/\gamma$, or to $-\infty$ if $u^2 < a^2/\gamma$, which explains the source of the corresponding separation of cases in the example of Section 4.

The fact that there are four singular points is a direct consequence of the nature of the F-functions on the right sides of Equations 5.13 - 5.17. Consider the set of equations obtained by equating all the F's to zero. If one is interested only in locating the singular points (not in the null surfaces as such), it is perfectly permissible to substitute from one equation $F = 0$ to another, still keeping five equations for five surfaces, the common intersections of which are the singular points. Thus, Equation 5.14 can be used to eliminate v from Equations 5.15 and 5.17 and Equation 5.16 ($J_z = 0$) can be used to eliminate $J_z$ from Equation 5.17. Note that the resulting equations are linear in v, T, and $J_z$ and quadratic in $\tau$ and $B_y$ (if the null surface of Equation 5.13 is multiplied by $\tau$). But a set of linear equations can produce at most only one singular point; i.e., they are equations of planes, and three or more planes can obviously have no more than one common point. A single quadratic equation combined with the linear equations can produce at most two singular points, since a parabola and a straight line can have at most two intersections. If there are two quadratic equations among the set of F's, each can have two zeros, with the result that the combined system can have at most four singular points; i.e., two parabolas with nonparallel axes can have at most four intersections.

It is now possible to conclude that the existence and uniqueness properties of the full five-dimensional system are the same as in the three limiting cases. The primary reasons for arriving at this conclusion are:

1. All three limit cases give topologically identical results.
2. The limits are attained in a continuous manner.
3. The two configurations of integral curves found are the only ones which are compatible with all of the known facts related to the behavior of the integral curves.

## Chapter 7

# RESULTS AND CONCLUSIONS

We have studied the properties of plane shock waves in an ionized gas in the presence of an external magnetic field oriented at an arbitrary angle with respect to the plane of the shock. These properties were analyzed in the temperature-density-field region in which the cyclotron frequency of the electrons is small compared with the collision frequencies, and in which relativistic effects, radiation pressure, and radiation energy density can be neglected. In addition, we assumed that equilibrium between ions and electrons is re-established downstream of the shock.

Chapter 2 reviewed and extended the theory of steady-plane shock waves considered as discontinuities in a nondissipative fluid. For fixed values of the mass flux, normal- and transverse-momentum flux, stagnation enthalpy, and normal magnetic field, we showed that the end states of shocks are conveniently represented geometrically as the points of intersection of three surfaces in a three-dimensional space, in which the coordinates are temperature, specific volume, and the component of the magnetic field parallel to the shock surface. Since one of these surfaces changes scale as a function of normal-momentum flux, another as a function of transverse-momentum flux, and the third as a function of stagnation enthalpy, it is easy to visualize how the end states vary as these constants vary. A second useful representation of magnetohydrodynamic shocks is in terms of the shock adiabatic. For the case of an ideal gas it was found that in the plane of the Alfvén numbers of the flow upstream and downstream of the shock, all finite amplitude shocks moving into a fluid which has given ratios of the fast and slow sound speeds to the normal Alfvén speed lie on a single S-shaped curve. The equations for this curve and also for the normalized speeds of fast and slow sound behind the shock were found in a form very convenient for further theoretical analysis or for computation.

In Chapter 3 we considered the stability of shock waves by allowing small-flow perturbations about a steady shock wave idealized as a discontinuity in a nondissipative fluid. This study has been chiefly an elaboration and clarification of the work of several Russian authors mentioned in the text, and the main results are those reported by them: namely, that only fast-and slow-shock waves can be stable; intermediate shocks and the shocks which form the boundaries of the intermediate-shock region with the fast-and slow-shock regions, "switch-on" and "switch-off" shocks, respectively, appear to disintegrate in such a way that the growth of the disturbances from infinitesimal size cannot be followed by a linear analysis. To obtain further insight into this instability, we studied the reflection and refraction of small normal Alfvén waves incident on shocks in the fast and slow regions. We found that the shock wave adds energy flux to the diverging waves, and that this energy flux becomes unbounded as the boundary with the intermediate-shock region is approached. We interpret this to mean that at the boundary all of the energy of the shock goes into the diverging waves.

In Chapter 4, a study of the steady-state one-dimensional shock layer began by deriving, from kinetic theory, a set of macroscopic equations valid for shocks. The principal conclusion obtained during this derivation was that the ordinary macroscopic equations of magnetohydrodynamics are not valid for shocks in that: (1) the current-inertia terms in the generalized Ohm's law are at least as important as the collision term (term proportional to electrical resistivity), and (2) the electric pressure inside the shock can be of the order of magnetic and gas pressures. A set of seven first-order ordinary nonlinear differential equations for the steady-state shock layer was found.

Chapters 5 and 6 examined the problems of existence and uniqueness of the shock layer and of its qualitative behavior. First, it was found that coupling from the electric-pressure term in the normal-momentum equation does not affect the existence and uniqueness properties of shocks, with the result that a system of only five differential equations need be considered if only those properties are being sought. (The fifth-order system differs from the system analyzed by Germain[13] in that current-inertia effects are included.) The main conclusions of this study are that unique fast shocks always exist; a unique shock-layer curve for slow shocks always exists, but in some extreme cases it may intersect the plane $\tau = 0$; unique $1 \to 3$ and $2 \to 4$ shocks and an infinite family of $1 \to 4$ shocks exist when the first viscosity and thermal conductivity are small compared with the second viscosity, electrical resistivity, and current inertia; a unique $2 \to 3$ shock-layer curve exists for one particular set of values of the dissipation coefficients; and there are no intermediate shocks when the first viscosity is large. In the limiting case corresponding to "switch-on" and "switch-off" shocks, the same conclusions apply, namely that unique shock-layer curves corresponding to these two shocks exist, but there is no shock layer corresponding to the $2 \to 3$ shock, which has now degenerated into a rotational or Alfvén discontinuity. (Landau and Lifshitz[1] also show that this discontinuity cannot exist in the steady state.) The arguments given indicate that only the fast and slow shocks exist independently of the values of the dissipation coefficients. Intermediate shocks, which have been shown to be unstable with respect to small disturbances, possess steady-state structure only for certain ranges of values of the dissipation coefficients.

The qualitative structure of the shock layer has also become clear from the analysis of Chapters 5 and 6. Two types of spatial oscillations may arise within the shock layer: those which are due to the electric field, and those due to current inertia. The properties of the former are apparent from Equation 4.80, which shows that an observer at rest with respect to the fluid ahead of the shock will see oscillations at the plasma frequency $\omega_p$ if the damping is below critical. Inside the shock, the damping may be zero or even negative; however, as the normal-velocity gradient subsides, the damping becomes proportional to the collision frequency $\nu_c$, and outside the shock, oscillations will appear only if the plasma frequency is greater than half the collision frequency. The importance of collisions can therefore be assessed qualitatively by observing the amount of radiation from the shock at the plasma frequency. Oscillations due to current inertia appear downstream of the shock whenever the determinant of Equations 5.25 has a pair of complex roots, in which case the frequency and damping of those oscillations are determined by the imaginary and real parts of those roots, respectively. In the case in which viscosity and thermal conductivity are small, a simpler criterion for the appearance of current-inertia oscillations is obtained from Equation 6.8, which shows that those oscillations can appear only if $\omega_p/\nu_c$ is of the order c/u. The damping of these oscillations is proportional to $\nu_c$ even inside the shock; i.e., there is no reduction of damping as in the case of electric-field oscillations. The physical "stiffness" which combines with electron inertia to produce these oscillations is the tangential magnetic field. When there are no oscillations, the temperature, density, and magnetic field vary monotonically through fast and slow shocks, and sometimes through $1 \to 3$, $2 \to 4$, and $3 \to 4$ shocks.

As a quantitative estimate of shock thickness, the formula $(\lambda_A)^{-1} + (-\lambda_B)^{-1}$ is suggested. The integral curve corresponding to the shock layer leaves the upstream singular point in the direction of the eigenvector corresponding to $\lambda_A$, and arrives at the downstream singular point in the direction of the eigenvector corresponding to $\lambda_B$ if $\lambda_B$ is real, or in the plane defined by the complex-conjugate pair of eigenvectors if $\lambda_B$ is complex. If $\lambda_B$ is complex, its real part appears in the above estimate of shock thickness. Since knowledge of the particular eigenvectors corresponding to the shock layer is a by-product of the existence and uniqueness proofs, the particular values $\lambda_A$ and $\lambda_B$ can always be selected from the five eigenvalues of Equation 5.25. It is interesting to note that $\lambda_A$ is the smallest of the positive eigenvalues, and $\lambda_B$ is algebraically the largest of the negative eigenvalues; hence, the number $(\lambda_A)^{-1} + (-\lambda_B)^{-1}$ is the largest that can be obtained by combining a positive and negative eigenvalue in this way.

Upon comparing the results of Chapter 3 with the results of Chapters 5 and 6, it is evident that fast and slow shocks are the only ones which have a steady-state structure and are also stable with respect to disintegration resulting from small disturbances in the flow. Intermediate shocks and the boundary cases of "switch-on" and "switch-off" shocks sometimes possess structure but must disintegrate in the presence of small disturbances. Included in the latter category is the normal shock for which the upstream flow velocity is super-Alfvénic and the downstream velocity sub-Alfvénic. These results clearly are not the same as those found in ordinary hydrodynamics, in which all shocks that possess a unique structure are also stable with respect to small disturbances in the flow.

Based upon the results of this study are the following suggestions for further work:

1. A normal shock which is super-Alfvénic upstream and sub-Alfvénic downstream is both mathematically and physically the simplest example of a shock which possesses steady-state structure but is not stable with respect to all small disturbances. Also, for this case, it does not appear possible to suggest a mode of breakup into a series of stable shocks as is possible for oblique intermediate shocks. It would seem, therefore, that to obtain an understanding of the instability process in this case it would be necessary to solve a stability problem in which the steady-state flow condition is the calculated shock-layer curve, including electric-field effects.

2. Shock-tube experiments should be conducted in the temperature-density-field region in which the present theory is applicable. Besides confirming the existence of the stable shocks, these experiments should be designed to provide quantitative data on shock thickness and on the properties of spatial oscillations in the flow, to be used for comparison with corresponding quantitative calculations from the present theory. Most of these quantitative calculations involve nothing worse than solution of a fifth-order polynomial; it does not appear necessary to solve the nonlinear shock-layer equations. It is also suggested that the phenomenon discussed in the first recommendation above be studied experimentally.

For that purpose, a normal magnetic field could be varied from the region in which the flow is sub-Alfvénic both upstream and downstream of the shock into the region in which the upstream flow is super-Alfvénic. At the transition point, the flow should undergo a marked change if the present theory is correct, and then when the normal field is large enough so that the flow becomes super-Alfvénic both upstream and downstream, the usual gas-dynamic shock structure should reappear.

3. The present theory was simplified into tractable form by restriction to the temperature-density-magnetic-field region in which the effects of orbiting particles, of radiation pressure, and of relativity could be neglected. Furthermore, the Navier-Stokes approximation was used to express the pressure tensor and heat-flux vector in terms of other variables. Since shock waves for which these assumptions no longer apply are of great interest, we suggest that the results of the present study on existence, uniqueness, and qualitative properties of shock waves be extended into a wider region of applicability. By building on the results of Chapters 5 and 6, it seems likely that this problem — hopelessly complicated to the author at the beginning of the present study — can be treated. (For example, it can be quite easily shown that inclusion of the primary Hall current term of Equation 4.58 does not affect the results on existence and uniqueness.) One of the main difficulties with respect to a more exact kinetic-theory discription is that of finding the correct expressions for the various components of the pressure and conductivity tensors, but thus far existence and uniqueness proofs have been obtainable without this knowledge.

APPENDIX

by William H. Heiser

## 1. Nomenclature Used in the Appendix

Physical Quantities:

| | |
|---|---|
| $b_x$ | Alfvén speed normal to shock front = $B_x / \sqrt{\rho\mu_0}$ |
| B | magnetic flux density |
| F | momentum flux density |
| G | mass flux density |
| H | stagnation enthalpy |
| k | ratio of specific heats |
| M | Mach number |
| n | particle density |
| $n_e$ | electron density |
| p | pressure |
| T | temperature |
| u | flow velocity normal to shock front |
| $u_0$ | flow constant = $B_x^2 / \mu_0 G$ |
| $Y_s$ | experimental bremsstrahlung spectral radiant intensity |
| $Y_{s0}$ | theoretical bremsstrahlung spectral radiant intensity for zero axial field |
| $Y_{s1}$ | theoretical bremsstrahlung spectral radiant intensity for nonzero axial field |
| $\lambda$ | wavelength |
| $\mu_0$ | magnetic permeability of free space |
| $\rho$ | density |

Subscripts:

| | |
|---|---|
| 1, 2, 3, 4 | stationary shock states(in order of increasing entropy) |
| x, y, z | Cartesian coodinate system axes (the shock being in the y-z plane) |
| I, II | wavelengths |

## 2. Introduction

The theory developedby Anderson does not completely resolve the question of which shock transition, if any, will be observed when the switch-on configuration is possible (that is, $B_{y1} = 0$, $B_x \neq 0$, and $u_1 \geqq u_0 \geqq u_4$). One could tentatively conclude that the switch-on transition (1-2 or 1-3) would occur in nature because it might be stable in the evolutionary sense, whereas the gasdynamic transition (1-4) is certainly not stable in the evolutionary sense. However, the switch-on transition was shown to be unstable to a fundamental type of infinitesimal disturbance, and, since the entropy increase is at a maximum across the gasdynamic shock wave, the perversity of nature would seem to require the 1-4 transition to occur. The reader is particularly referred to pages 61-68 for the reasoning that led Anderson to these conclusions. However, it should be kept in mind that Anderson did not strictly conclude that switch-on transitions were nonevolutionary, but only that they formed the boundary between evolutionary and nonevolutionary transitions. He then classified switch-on transitions as nonevolutionary because the equilibrium value of the amplitude of emergent waves was infinite when the shock layer was disturbed by a monochromatic wave train of finite amplitude and infinite extent. Since the occurrence of such a wave train is unlikely in any real experiment, it is still possible that the switch-on shock exists and can be produced experimentally.

In order to shed some light upon these aspects of shock-wave theory, experiments were conducted upon shock waves approaching the conditions of the foregoing theory by the Plasma Magnetogasdynamics Group, Research Laboratory of Electronics, * M. I. T. The following presentation contains some of the important results of this research.[1]

The coordinate system and nomenclature used in this appendix are chosen to be consistent with those of Anderson, and the MKS

* This research was supported in part by the U. S. Army Signal Corps, the Air Force Office of Scientific Research, and the Office of Naval Research; and in part by the National Science Foundation (Grant G-9330).

system is employed, except that some computed values are stated in more convenient or conventional units.

3. Description of the Experiments

Magnetically driven shock waves were produced in a shock tube of annular geometry similar to that used and described by Patrick[2] (Figure A. 1). Hydrogen was chosen as the experimental gas to ensure complete ionization of the shock-heated test gas, and the initial experimental gas pressure ($p_1$) was varied between 80 and 180 microns. When the drive current reached its maximum value of $2 \times 10^5$ amperes, the magnetic pressure in the void behind the expansion wave had a mean value of 0.7 $w/m^2$, corresponding to a pressure of about 2 atmospheres. The drive current remained fairly constant for about 3 μs, during which period all data were taken. The resulting shock velocities measured in laboratory coordinates (corresponding to $u_1$ of the theory) were approximately 8 cm / μs, which is equivalent to $M_1 > 60$ based upon room temperature. It is noted that the test gas temperature for a gasdynamic shock of this strength is of the order of $10^5$ °K.

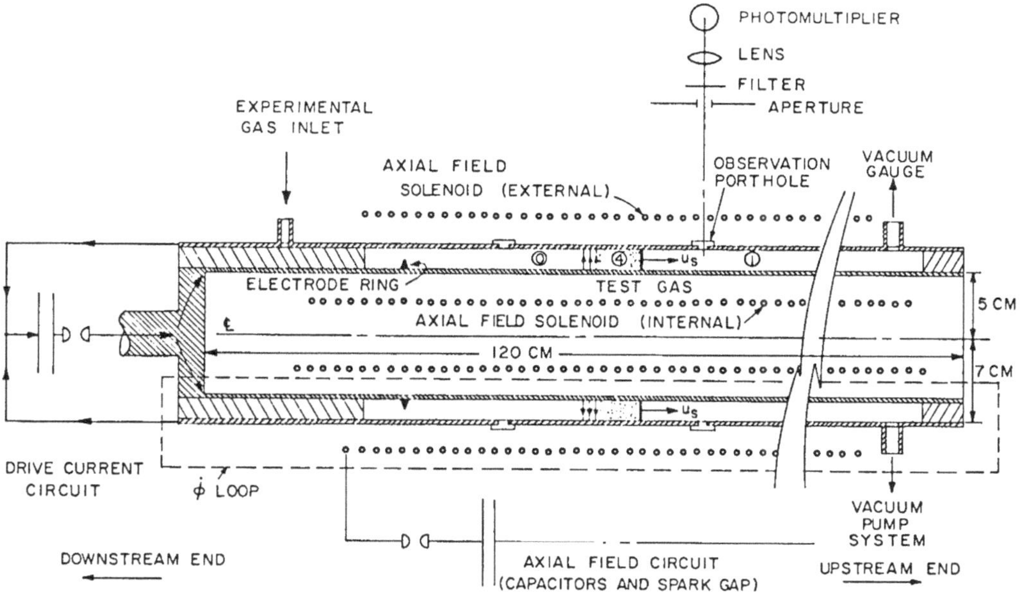


Figure A. 1. Schematic representation of a shock tube and auxiliary equipment

An axial magnetic field (Bx, normal to the shock front) that remained essentially constant during the lifetime of the shock wave was produced by means of an external and internal solenoid (Figure A. 1). Prior to each experiment the axial magnetic field strength was chosen from the following values ($w/m^2$): 0, 0.017, 0.034, 0.068, 0.102, 0.136, 0.170, 0.204, 0.238. Since the maximum resulting Alfvén speed in the experimental gas was about 6 cm/μs, the shock velocity always exceeded all small-wave velocities there, and the unshocked gas must have been in stationary state 1 of the theory.

Both the shock velocity and the density of the test gas were obtained by means of two sensitive, rapidly responding, calibrated photomultipliers. The photomultipliers were provided with optical systems that collected the light emitted by a known volume of test gas while filtering out all emitted light except that within a narrow band ( < 160 Å wide). They centered at known wavelengths ( $\lambda_I$ = 3800 Å, $\lambda_{II}$ = 5000 Å). The photomultipliers were periodically calibrated against a standard light source to allow the magnitude of the experimental signals to be converted to absolute spectral radiant intensity.

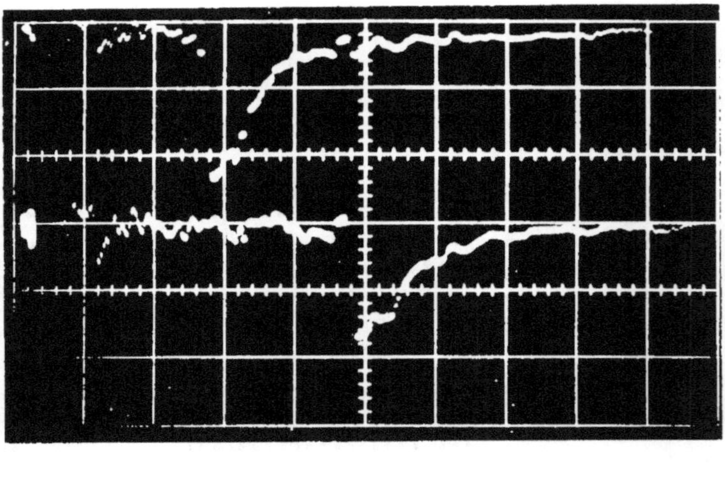

Figure A. 2. Data oscillogram, $B_x$ = 0.034 $w/m^2$, p = 136μ$H_2$

The oscillogram shown in Figure A. 2 simultaneously recorded as a function of time the light intensity observed by two separate photomultipliers located 15 cm apart along the axis of the shock tube. The sweep speed was 1 μs/cm; a downward deflection indicates an increase of light reaching the photomultipliers. The discontinuity of light intensity is caused by the arrival of the shock front, and the shock velocity ($u_1$) is computed directly from the time required for the shock to traverse the distance between the photomultipliers. The emitted light is largely due to continuum radiation, or bremsstrahlung, from the hot test gas, for which theoretical values may be obtained.

## 4. Theory

Continuum Radiation. Computed values of continuum spectral radiant intensity ($Y_{s0}$) divided by electron density ($n_e$) squared for hydrogen at a wavelength of 3800 Å in the absence of a magnetic field are presented as a function of temperature (T) in Figure A. 3, the Oster classical range calculation being strictly valid only for temperatures between $2 \times 10^5$ K and $9 \times 10^5$ K, and the quantum range calculation being valid for temperatures in excess of $9 \times 10^5$ K.[3,4] These results make it clear that $Y_{s0}$ is only a weak function of T for the anticipated experimental values of T, and is therefore closely proportional to $n_e^2$. Since the shock-heated hydrogen is completely ionized, the electron density is equal to the particle density (n).

Shock Algebra. Since $M_1 >> 1$ (i.e., $\rho_1 u_1^2 >> p_1$) and $B_{y1} = 0$, it follows that:

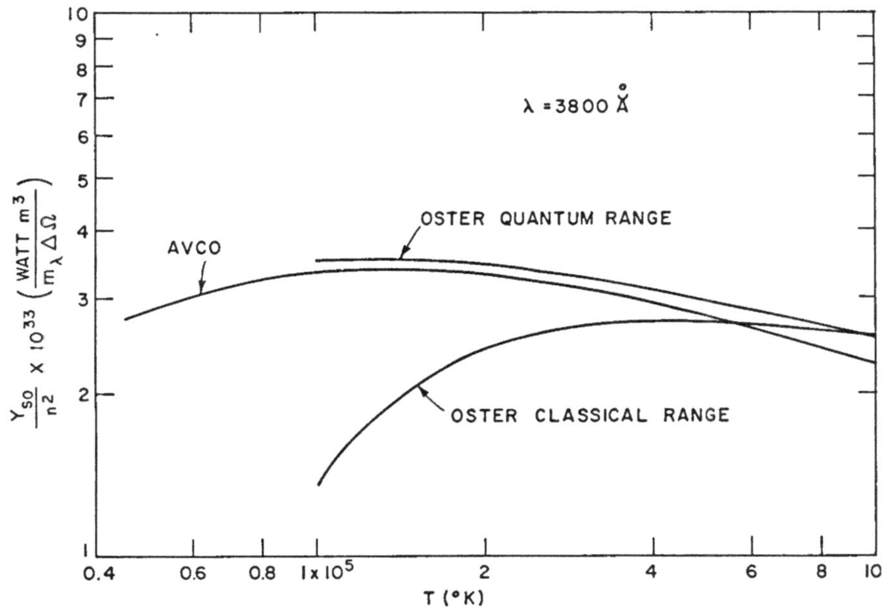


Figure A. 3. $Y_{s0}/n^2$ as a function of T, λ = 3800 Å

$$G = \rho_1 u_1,$$

$$F_x = \rho_1 u_1^2$$

$$F_y = 0$$

$$H = \frac{u_1^2}{2}$$

where $\rho_1$ and $u_1$ are determined experimentally.

Consequently, the conservation equations may be solved for the possible states downstream of the shock. The following solutions are easily obtained, assuming evolutionary theory to be correct:

a. Fast or gasdynamic shock (1-4) $u_4 \geqq u_0$:

$$B_{y4} = 0$$

$$\rho_4 = \left(\frac{k+1}{k-1}\right) \rho_1 = 4\rho_1$$

b. Intermediate or switch-on shock (1 - 2 or 1 - 3) $u_1 \geqq u_0 \geqq u_4$:

$$u_2 = u_0 = b_{x2}$$

$$\rho_2 = \left(\frac{u_1}{b_{x1}}\right)^2 \rho_1$$

$$\left(\frac{B_{y2}}{B_x}\right)^2 = (k-1)\left(\frac{u_1}{u_0} - 1\right)\left(\frac{k+1}{k-1} - \frac{u_1}{u_0}\right)$$

where transition occurs when

$$u_4 = u_0$$

or

$$u_1 = b_{x1}\sqrt{\frac{k+1}{k-1}} = 2b_{x1}$$

Since $Y_{s0} \propto n^2$, then for a given value of $\rho_1$,

$$\frac{Y_{s1}}{Y_{s0}} = 1 \text{ for } u_1 \geqq 2b_{x1}$$

and

$$\frac{Y_{s1}}{Y_{s0}} = \frac{1}{16}\left(\frac{\rho_2}{\rho_1}\right)^2 = \frac{1}{16}\left(\frac{u_1}{b_{x1}}\right)^4 \text{ for } 2b_{x1} \geqq u_1 \geqq b_{x1}$$

where $Y_{s1}$ is the theoretical value of continuum radiation for $B_x \neq 0$.

The change in the functional dependence of $Y_{s1}/Y_{s0}$ upon $u_1/b_{x1}$ at the point $u_1 = 2b_{x1}$ is fortuitously large. It is noted that both $u_1$ and $b_{x1}$ are experimentally measured quantities.

## 5. Experimental Results

Experimental verification of the predicted dependence of the measured spectral radiant intensity ($Y_s$) upon $p_1^2$ for $\lambda = 3800$ Å and moderate values of $B_x(u_1 > 2b_{x1})$ is presented in Figure A.4,

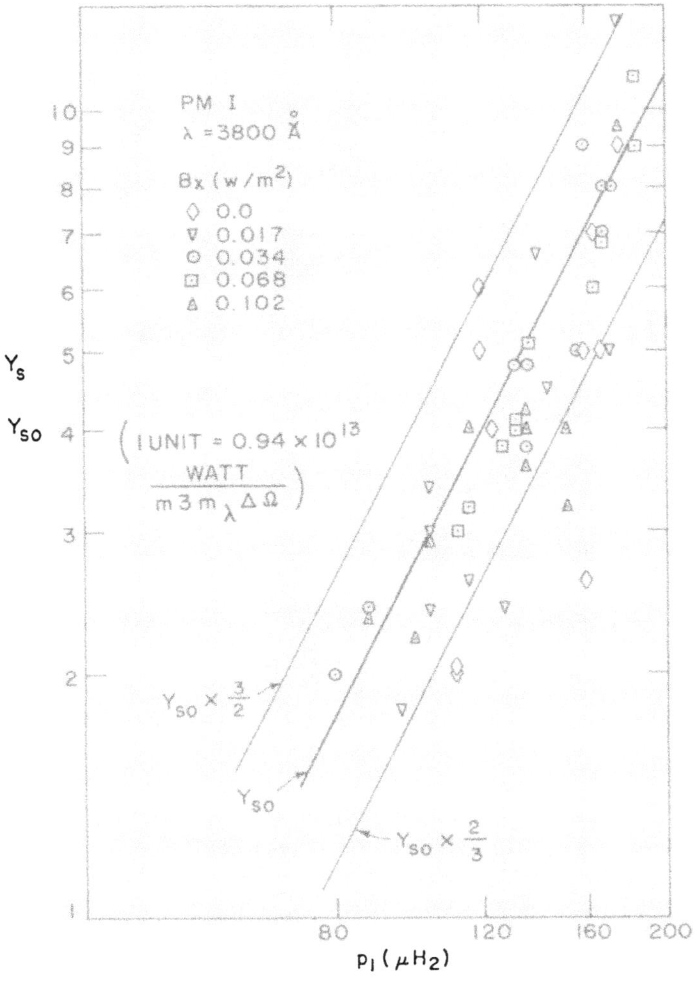


Figure A.4. $Y_s$ and $Y_{s0}$ as functions of $B_x$ and $p_1$, $B_x \leqq 0.102$ w/m$^2$, $\lambda = 3800$ Å

where $Y_{s0}$ was based upon the assumptions that $\rho_4 = 4\rho_1$ and $T_4 = 10^5$ K. The excellent agreement between theory and experiment is taken as an indication that the latter assumptions were reasonable.

The predicted behavior of $Y_{s1}/Y_{s0}$ and the observed behavior of $Y_s/Y_{s0}$ as functions of $b_{x1}/u_1$ are shown in Figure A.5 (where $u_s$ corresponds to $u_1$), and the data for large values of $B_x$ are separated for the ease of interpretation in Figure A.6. Because a break in the data occurred near the predicted value of $u_1 = 2b_{x1}$, and because the value $Y_s/Y_{s0}$ decreased almost as rapidly with $b_{x1}/u_1$ as predicted for $2b_{x1} \geqq u_1$, these results strongly suggest that the switch-on shock was observed for $u_1 \geqq u_0 \geqq u_4$.

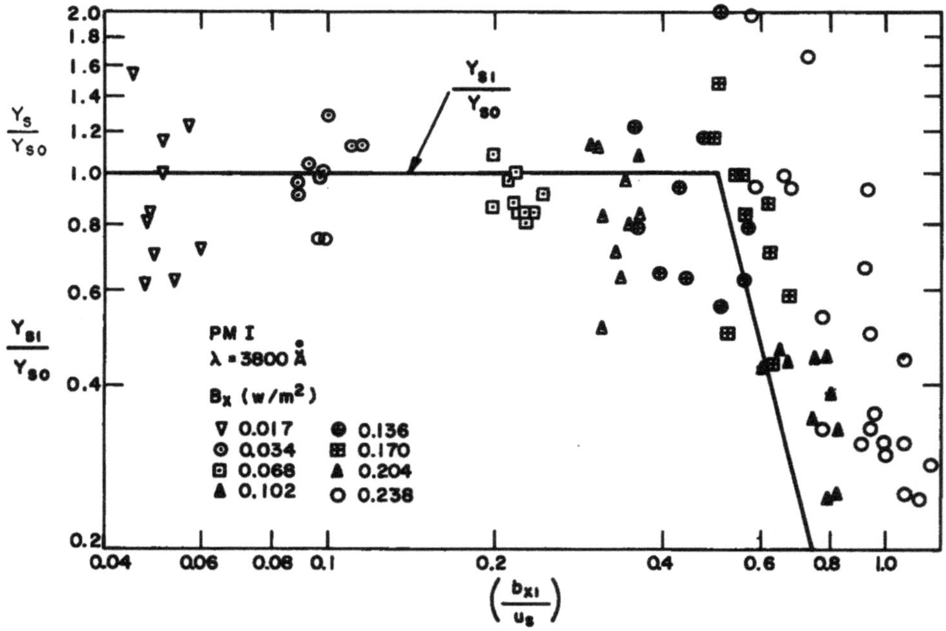


**Figure A.5. $Y_s/Y_{s0}$ and $Y_{s1}/Y_{s0}$ as functions of $b_{x1}/u_s = b_{x1}/u_1$, $\lambda$ = 3800 Å**

As is often the case with experiments employing current arcs, not every attempt was successful in producing a recognizable shock as recorded by the photomultipliers. This might be interpreted as an indication of the instabilities of the shock layer predicted by Anderson. However, the probability of producing a shock upon any given attempt was not a strong function of axial magnetic field, even when this field was small or zero, which is a sign that the instability mechanism was located elsewhere, the probable location being the drive-current expansion wave.

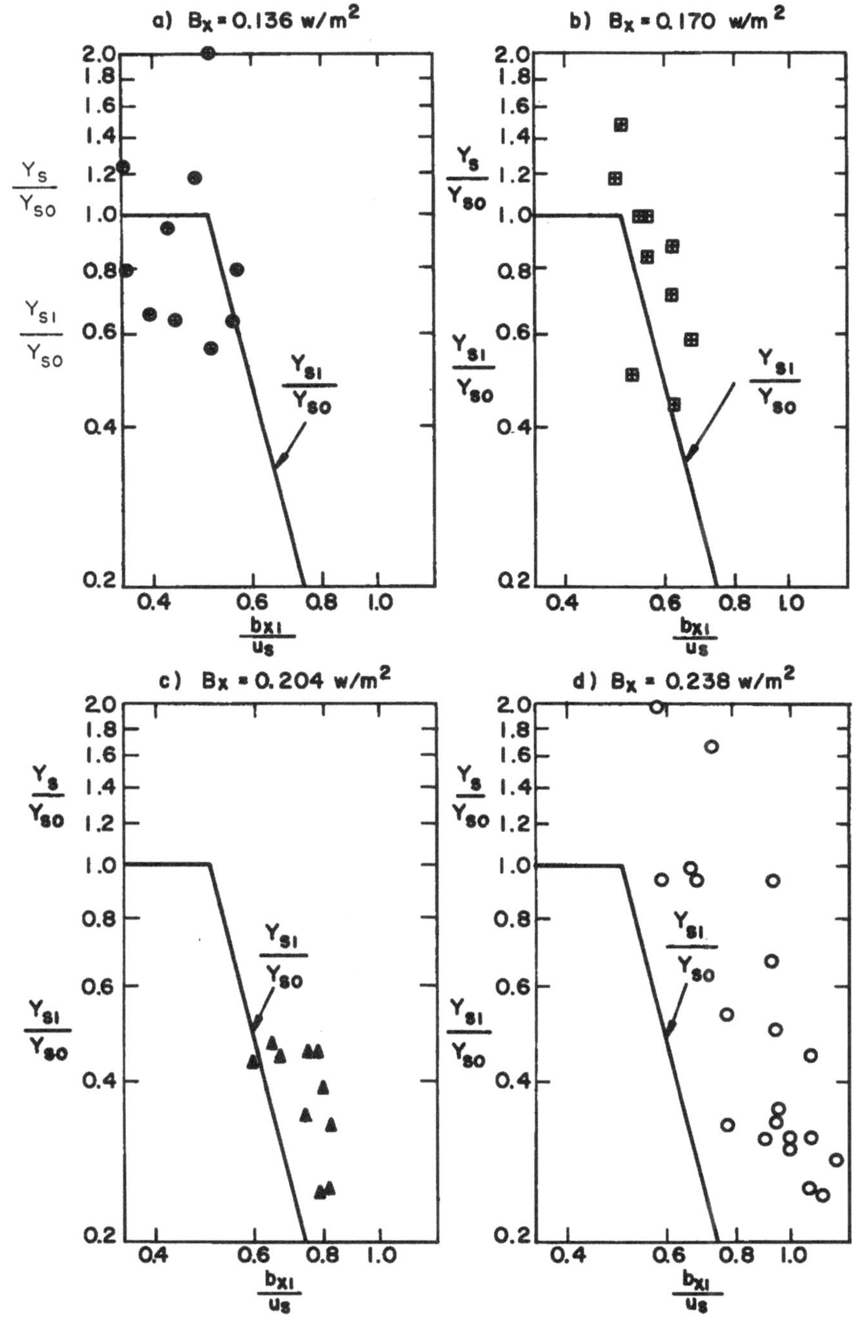


Figure A.6. $Y_s/Y_{s0}$ and $Y_{s1}/Y_{s0}$ as functions of $b_{x1}/u_s = b_{x1}/u_1$, and $B_x$. $B_x \geqq 0.136\ w/m^2$, $\lambda = 3800\ \text{Å}$

## 6. Conclusions

The experimental results prove that some form of shock wave exists when $u_1 \geqq u_0 \geqq u_4$, and comparison with theory indicates that the stable transition in this case is probably the switch-on shock. For $u_4 \geqq u_0$, the experiments reveal that the gasdynamic shock is the stable transition. These results verify evolutionary shock theory.

Two experiments immediately suggest themselves. The first would require an increase of the axial magnetic field to make $u_0 \geqq u_1$, for which case transition algebra predicts a return of the gasdynamic shock. However, the flow downstream of the shock would be sub-Alfvénic, allowing the drive current to overtake the shock front, with consequences not included in the transition algebra. The second would repeat these experiments in a gas-driven shock tube. The currents that flowed in the shock front to produce $B_{y2}$ in the experiments reported here were taken from the drive current and were carried to the shock front by the slow magnetoacoustic wave. Since drive current does not exist in a gas-driven shock tube, circulating currents would be required to produce the switch-on transition, which would be a severer test of evolutionary theory.

## Appendix References

1. W. H. Heiser, "Axial Field Effects in a Magnetically Driven Shock Tube," M. I. T. Ph. D. thesis (Sept. 1962).

2. R. M. Patrick, "The Production and Study of High Speed Shock Waves in a Magnetic Annular Shock Tube," AVCO Research Report No. 59 (July 1959).

3. G. S. Janes and H. E. Koritz, "Numerical Calculation of Absolute Bremsstrahlung Intensity for a Fully Ionized Fully Dissociated Hydrogenic Gas," AVCO Research Report No. 70 (Sept. 1959).

4. L. Oster, "Emission, Absorption, and Conductivity of a Fully Ionized Gas at Radio Frequencies," Revs. Modern Phys. 33, No. 4 (Oct. 1961)., p. 525.

## REFERENCES


1. L. D. Landau and E. M. Lifshitz, Electrodynamics of Continuous Media, Addison-Wesley Publishing Co., Reading, Mass. (1960).
2. F. de Hoffmann and E. Teller, "Magneto-Hydrodynamic Shocks," Phys. Rev. 80, 692 (1950).
3. K. O. Friedrichs, "Nonlinear Wave Motion in Magneto-Hydrodynamics" (Los Alamos Report No. LAMS-2105 (Physics); written Sept. 1954, distributed Mar. 8, 1957.
4. H. L. Helfer, "Magneto-Hydrodynamic Shock Waves," Astrophys. J. 117, 177 (1953).
5. R. Lüst, "Magneto-hydrodynamische Stosswellen in einem Plasma unendlicher Leitfähigkeit," Z. Naturforsch. 8a, 277 (1953).
6. R. Lüst, "Stationäre magneto-hydrodynamische Stosswellen beliebiger Stärke," Z. Naturforsch. 10a, 125 (1955).
7. J. Bazer and W. B. Ericson, "Hydromagnetic Shocks," Astrophys. J. 129, 758 (1959).
8. L. Napolitano, "Discontinuity Surfaces in Magnetofluid-Dynamics," Polytechnic Institute of Brooklyn Aeronautical Laboratory Report No. 503, Dec. 1958.
9. W. D. Hayes, Gasdynamic Discontinuities, Princeton University Press, Princeton, N. J. (1960).
10. D. Gilbarg, "The Existence and Limit Behavior of the One-Dimensional Shock Layer," Am. J. Math. 73, 256 (1951).
11. D. Gilbarg and D. Paolucci, "The Structure of Shock Waves in the Continuum Theory of Fluids," J. Rat. Mech. Anal. 2, 617 (1953).
12. J. M. Burgers, "On the Transmission of Sound Waves through a Shock Wave," Koninklijke Nederlandsche Akademie van Wetenschappen, Proceedings, 49, 273 (1946).
13. P. Germain, "Contribution à la Théorie des Ondes de Choc en Magnétodynamique des Fluides," Office National d'Études et de Recherches Aéronautiques, Publication No. 97, Dec. 1959. See also Revs. Modern Phys. 32, 951 (1960).
14. J. A. Shercliff, "One-Dimensional Magnetogasdynamics in Oblique Fields," J. Fluid Mech. 9, 481 (1960).

15. H. Weyl, "Shock Waves in Arbitrary Fluids," Communs. Pure and Appl. Math. 2, 103 (1949).

16. R. V. Polovin and V. P. Demutskii, "Udarna Adiabata v Magnitnoi Gidrodinamike," Ukrain. Fiz. Zhur. 5, 3 (1960).

17. R. V. Polovin, "Shock Waves in Magnetohydrodynamics," Soviet Phys. Uspekhi 3, 677 (1961).

18. L. D. Landau and E. M. Lifshitz, Fluid Mechanics, Addison-Wesley Publishing Co., Reading, Mass. (1959).

19. N. E. Kotchine, "Sur la Théorie des Ondes de Choc dans un Fluide," Rendiconti del Circlo Matematico di Palermo 50, 305 (1926).

20. V. V. Gogosov, "Resolution of an Arbitrary Discontinuity in Magnetohydrodynamics," Applied Math. and Mech. (PMM) 25, 108 (1961).

21. V. V. Gogosov, "Interaction of Magnetohydrodynamic Waves with Contact and Vortex Discontinuities," Applied Math. and Mech. (PMM) 25, 187 (1961).

22. V. V. Gogosov, "Bzaimodeistvie Magnitogidrodinamicheskikh Voln," Prikladnaia Matematika i Mekhanika 25, 456 (1961).

23. V. M. Kontorovich, "On the Interaction Between Small Disturbances and Discontinuities in Magnetohydrodynamics and the Stability of Shock Waves," Soviet Phys.-JETP 8, 851 (1959).

24. G. I. Liubarskii and R. V. Polovin, "The Disintegration of Unstable Shock Waves in Magnetohydrodynamics," Soviet Phys.-JETP 9, 902 (1959).

25. H. and B. S. Jeffreys, Methods of Mathematical Physics, 3rd ed., Cambridge at the University Press (1956), Chap. 3.

26. Boa-Teh Chu, "Thermodynamics of Electrically Conducting Fluids," Physics of Fluids 2, 473 (1959).

27. S. I. Syrovatskii, "Magnitnoigidrodinamike," Uspekhi Fiz. Nauk 62, 247 (1957).

28. G. A. Sawyer, P. L. Scott, and T. F. Stratton, "Experimental Demonstration of Hydromagnetic Waves in an Ionized Gas," Physics of Fluids 2, 47 (1959).

29. M. J. Lighthill, "Studies of Magneto-Hydrodynamic Waves and Other Anisotropic Wave Motions," Phil. Trans. Roy. Soc. London 252, 397 (1960).

30. T. G. Cowling, Magnetohydrodynamics, Interscience Publishers, New York (1957).

31. H. Grad in D. Bershader, ed., The Magnetodynamics of Conducting Fluids, Stanford University Press, Stanford, Calif. (1959); Symposium on Magnetohydrodynamics, 3rd, Palo Alto, Calif., 1958.

32. W. R. Sears, "Some Remarks about Flow past Bodies," Revs. Modern Phys. 32, 701 (1960).

33. E. L. Ince, Ordinary Differential Equations, Dover Publications, New York (1956), Sec. 1.3.

34. A. I. Akhiezer, G. I. Liubarskii, and R. V. Polovin, "The Stability of Shock Waves in Magnetohydrodynamics," Soviet Phys.-JETP 8, 507 (1959).

35. S. I. Syrovatskii, "The Stability of Shock Waves in Magnetohydrodynamics," Soviet Phys.-JETP 8, 1024 (1959).

36. H. Grad, comment in discussion, Revs. Modern Phys. 32, 958 (1960).

37. H. Grad, "The Profile of a Steady Plane Shock Wave," Communs. Pure and Appl. Math. 5, 257 (1952).

38. L. Talbot and F. S. Sherman, "Structure of Weak Shock Waves in a Monatomic Gas," NASA Memo 12-14-58W, Jan. 1959.

39. S. Ziering, F. Ek, and P. Koch, "Two-Fluid Model for the Structure of Neutral Shock Waves," Physics of Fluids 4, 975 (1961).

40. J. K. Haviland, "Monte-Carlo Application to Molecular Flows," Ph.D. thesis, M.I.T., June 1961.

41. R. M. Patrick, "High-Speed Shock Waves in a Magnetic Annular Shock Tube," Physics of Fluids 2, 589 (1959).

42. C. S. Morawetz, "Magnetohydrodynamic-Shock Structure without Collisions," Institute of Mathematical Sciences, New York University, NYO-8677, Jan. 13, 1959.

43. W. Marshall, "The Structure of Magnetohydrodynamic Shock Waves," Proc. Roy. Soc. London A233, 367 (1955).

44. C. S. S. Ludford, "The Structure of a Hydromagnetic Shock in Steady Plane Motion," J. Fluid Mech. 5, 67 (1959).

45. A. G. Kulikovskii and G. A. Liubimov, "Structure of Oblique Magnetohydrodynamic Shock Waves," Applied Math. and Mech. (PMM) 25, 125 (1961).

46. H. S. Green, "Ionic Theory of Plasmas and Magnetohydrodynamics," Physics of Fluids 2, 341 (1959).

47. L. Spitzer, Physics of Fully Ionized Gases, Interscience Publishers, New York (1956).

48. J. L. Delcroix, Introduction to the Theory of Ionized Gases, Interscience Publishers, New York (1960).

49. J. O. Hirshfelder, C. F. Curtiss, and R. B. Bird, Molecular Theory of Gases and Liquids, John Wiley & Sons, New York (1954).

50. S. R. de Groot, Thermodynamics of Irreversible Processes, North-Holland Publishing Co., Amsterdam, and Interscience Publishers, New York (1951).

51. W. K. H. Panofsky and M. Phillips, Classical Electricity and Magnetism, Addison-Wesley Publishing Co., Reading, Mass. (1955).

52. S. Lefschetz, Differential Equations: Geometric Theory, Interscience Publishers, New York (1957).

53. N. McLachlan, Ordinary Non-Linear Differential Equations in Engineering and Physical Sciences," 2nd ed., Oxford at the Clarendon Press (1956).

54. R. von Mises, "On the Thickness of a Steady Shock Wave," J. Inst. Aeronautical Sci. 17, 551 (1950).

55. E. L. Ince, Ordinary Differential Equations, Dover Publications (1956); original edition by Longmans, Green & Co. (1926), Sec. 3.31.

56. H. Goldstein, Classical Mechanics, Addison-Wesley Publishing Co., Cambridge, Mass., (1950), Sec. 10-2.

57. H. and B. S. Jeffreys, Methods of Mathematical Physics, Cambridge at the University Press, 3rd ed. (1956), Chap. 4.

58. J. L. Synge, Relativity: the Special Theory," North-Holland Publishing Co., Amsterdam (1956), Sec. 1.15.

59. R. Courant and D. Hilbert, Methods of Mathematical Physics, Vol. I, Interscience Publishers, New York (1953), Chap. 1.

60. R. F. Scott, The Theory of Determinants and Their Applications, 2nd ed., Cambridge at the University Press (1904), p. 64.

61. N. Levinson, "Perturbations of Discontinuous Solutions of Non-Linear Systems of Differential Equations," Acta Mathematica 82, 71 (1950).

# INDEX

www.ingramcontent.com/pod-product-compliance
Lightning Source LLC
Chambersburg PA
CBHW021555210326
41599CB00010B/452
* 9 7 8 0 2 6 2 5 1 1 5 9 9 *